IET ENERGY ENGINEERING SERIES 239

Intelligent Control of Medium and High Power Converters

Other volumes in this series:

Volume 1 **Power Circuit Breaker Theory and Design** C.H. Flurscheim (Editor)
Volume 4 **Industrial Microwave Heating** A.C. Metaxas and R.J. Meredith
Volume 7 **Insulators for High Voltages** J.S.T. Looms
Volume 8 **Variable Frequency AC Motor Drive Systems** D. Finney
Volume 10 **SF_6 Switchgear** H.M. Ryan and G.R. Jones
Volume 11 **Conduction and Induction Heating** E.J. Davies
Volume 13 **Statistical Techniques for High Voltage Engineering** W. Hauschild and W. Mosch
Volume 14 **Uninterruptible Power Supplies** J. Platts and J.D. St Aubyn (Editors)
Volume 15 **Digital Protection for Power Systems** A.T. Johns and S.K. Salman
Volume 16 **Electricity Economics and Planning** T.W. Berrie
Volume 18 **Vacuum Switchgear** A. Greenwood
Volume 19 **Electrical Safety: A guide to causes and prevention of hazards** J. Maxwell Adams
Volume 21 **Electricity Distribution Network Design, 2nd Edition** E. Lakervi and E.J. Holmes
Volume 22 **Artificial Intelligence Techniques in Power Systems** K. Warwick, A.O. Ekwue and R. Aggarwal (Editors)
Volume 24 **Power System Commissioning and Maintenance Practice** K. Harker
Volume 25 **Engineers' Handbook of Industrial Microwave Heating** R.J. Meredith
Volume 26 **Small Electric Motors** H. Moczala *et al.*
Volume 27 **AC-DC Power System Analysis** J. Arrillaga and B.C. Smith
Volume 29 **High Voltage Direct Current Transmission, 2nd Edition** J. Arrillaga
Volume 30 **Flexible AC Transmission Systems (FACTS)** Y.-H. Song (Editor)
Volume 31 **Embedded Generation** N. Jenkins *et al.*
Volume 32 **High Voltage Engineering and Testing, 2nd Edition** H.M. Ryan (Editor)
Volume 33 **Overvoltage Protection of Low-Voltage Systems, Revised Edition** P. Hasse
Volume 36 **Voltage Quality in Electrical Power Systems** J. Schlabbach *et al.*
Volume 37 **Electrical Steels for Rotating Machines** P. Beckley
Volume 38 **The Electric Car: Development and future of battery, hybrid and fuel-cell cars** M. Westbrook
Volume 39 **Power Systems Electromagnetic Transients Simulation** J. Arrillaga and N. Watson
Volume 40 **Advances in High Voltage Engineering** M. Haddad and D. Warne
Volume 41 **Electrical Operation of Electrostatic Precipitators** K. Parker
Volume 43 **Thermal Power Plant Simulation and Control** D. Flynn
Volume 44 **Economic Evaluation of Projects in the Electricity Supply Industry** H. Khatib
Volume 45 **Propulsion Systems for Hybrid Vehicles** J. Miller
Volume 46 **Distribution Switchgear** S. Stewart
Volume 47 **Protection of Electricity Distribution Networks, 2nd Edition** J. Gers and E. Holmes
Volume 48 **Wood Pole Overhead Lines** B. Wareing
Volume 49 **Electric Fuses, 3rd Edition** A. Wright and G. Newbery
Volume 50 **Wind Power Integration: Connection and system operational aspects** B. Fox *et al.*
Volume 51 **Short Circuit Currents** J. Schlabbach
Volume 52 **Nuclear Power** J. Wood
Volume 53 **Condition Assessment of High Voltage Insulation in Power System Equipment** R.E. James and Q. Su
Volume 55 **Local Energy: Distributed generation of heat and power** J. Wood
Volume 56 **Condition Monitoring of Rotating Electrical Machines** P. Tavner, L. Ran, J. Penman and H. Sedding
Volume 57 **The Control Techniques Drives and Controls Handbook, 2nd Edition** B. Drury
Volume 58 **Lightning Protection** V. Cooray (Editor)
Volume 59 **Ultracapacitor Applications** J.M. Miller
Volume 62 **Lightning Electromagnetics** V. Cooray

Volume 63	**Energy Storage for Power Systems, 2nd Edition** A. Ter-Gazarian
Volume 65	**Protection of Electricity Distribution Networks, 3rd Edition** J. Gers
Volume 66	**High Voltage Engineering Testing, 3rd Edition** H. Ryan (Editor)
Volume 67	**Multicore Simulation of Power System Transients** F.M. Uriate
Volume 68	**Distribution System Analysis and Automation** J. Gers
Volume 69	**The Lightening Flash, 2nd Edition** V. Cooray (Editor)
Volume 70	**Economic Evaluation of Projects in the Electricity Supply Industry, 3rd Edition** H. Khatib
Volume 72	**Control Circuits in Power Electronics: Practical issues in design and implementation** M. Castilla (Editor)
Volume 73	**Wide Area Monitoring, Protection and Control Systems: The enabler for smarter grids** A. Vaccaro and A. Zobaa (Editors)
Volume 74	**Power Electronic Converters and Systems: Frontiers and applications** A.M. Trzynadlowski (Editor)
Volume 75	**Power Distribution Automation** B. Das (Editor)
Volume 76	**Power System Stability: Modelling, analysis and control** A.A. Sallam and B. Om P. Malik
Volume 78	**Numerical Analysis of Power System Transients and Dynamics** A. Ametani (Editor)
Volume 79	**Vehicle-to-Grid: Linking electric vehicles to the smart grid** J. Lu and J. Hossain (Editors)
Volume 81	**Cyber-Physical-Social Systems and Constructs in Electric Power Engineering** S. Suryanarayanan, R. Roche and T.M. Hansen (Editors)
Volume 82	**Periodic Control of Power Electronic Converters** F. Blaabjerg, K. Zhou, D. Wang and Y. Yang
Volume 86	**Advances in Power System Modelling, Control and Stability Analysis** F. Milano (Editor)
Volume 87	**Cogeneration: Technologies, optimisation and implementation** C. A. Frangopoulos (Editor)
Volume 88	**Smarter Energy: From smart metering to the smart grid** H. Sun, N. Hatziargyriou, H.V. Poor, L. Carpanini and M.A. Sánchez Fornié (Editors)
Volume 89	**Hydrogen Production, Separation and Purification for Energy** A. Basile, F. Dalena, J. Tong and T.N. Veziroğlu (Editors)
Volume 90	**Clean Energy Microgrids** S. Obara and J. Morel (Editors)
Volume 91	**Fuzzy Logic Control in Energy Systems with Design Applications in MATLAB®/SIMULINK®** İ.H. Altaş
Volume 92	**Power Quality in Future Electrical Power Systems** A.F. Zobaa and S.H.E. A. Aleem (Editors)
Volume 93	**Cogeneration and District Energy Systems: Modelling, analysis and optimization** M.A. Rosen and S. Koohi-Fayegh
Volume 94	**Introduction to the Smart Grid: Concepts, technologies and evolution** S.K. Salman
Volume 95	**Communication, Control and Security Challenges for the Smart Grid** S.M. Muyeen and S. Rahman (Editors)
Volume 96	**Industrial Power Systems with Distributed and Embedded Generation** R. Belu
Volume 97	**Synchronized Phasor Measurements for Smart Grids** M.J.B. Reddy and D.K. Mohanta (Editors)
Volume 98	**Large Scale Grid Integration of Renewable Energy Sources** A. Moreno-Munoz (Editor)
Volume 100	**Modeling and Dynamic Behaviour of Hydropower Plants** N. Kishor and J. Fraile-Ardanuy (Editors)
Volume 101	**Methane and Hydrogen for Energy Storage** R. Carriveau and D. S-K. Ting
Volume 104	**Power Transformer Condition Monitoring and Diagnosis** A. Abu-Siada (Editor)
Volume 106	**Surface Passivation of Industrial Crystalline Silicon Solar Cells** J. John (Editor)
Volume 107	**Bifacial Photovoltaics: Technology, applications and economics** J. Libal and R. Kopecek (Editors)

Volume 108 **Fault Diagnosis of Induction Motors** J. Faiz, V. Ghorbanian and G. Joksimović
Volume 109 **Cooling of Rotating Electrical Machines: Fundamentals, modelling, testing and design** D. Staton, E. Chong, S. Pickering and A. Boglietti
Volume 110 **High Voltage Power Network Construction** K. Harker
Volume 111 **Energy Storage at Different Voltage Levels: Technology, integration, and market aspects** A.F. Zobaa, P.F. Ribeiro, S.H.A. Aleem and S.N. Afifi (Editors)
Volume 112 **Wireless Power Transfer: Theory, technology and application** N. Shinohara
Volume 114 **Lightning-Induced Effects in Electrical and Telecommunication Systems** Y. Baba and V.A. Rakov
Volume 115 **DC Distribution Systems and Microgrids** T. Dragičević, F. Blaabjerg and P. Wheeler
Volume 116 **Modelling and Simulation of HVDC Transmission** M. Han (Editor)
Volume 117 **Structural Control and Fault Detection of Wind Turbine Systems** H.R. Karimi
Volume 118 **Modelling and Simulation of Complex Power Systems** A. Monti and A. Benigni
Volume 119 **Thermal Power Plant Control and Instrumentation: The control of boilers and HRSGs, 2nd Edition** D. Lindsley, J. Grist and D. Parker
Volume 120 **Fault Diagnosis for Robust Inverter Power Drives** A. Ginart (Editor)
Volume 121 **Monitoring and Control using Synchrophasors in Power Systems with Renewables** I. Kamwa and C. Lu (Editors)
Volume 123 **Power Systems Electromagnetic Transients Simulation, 2nd Edition** N. Watson and J. Arrillaga
Volume 124 **Power Market Transformation** B. Murray
Volume 125 **Wind Energy Modeling and Simulation Volume 1: Atmosphere and plant** P. Veers (Editor)
Volume 126 **Diagnosis and Fault Tolerance of Electrical Machines, Power Electronics and Drives** A.J.M. Cardoso
Volume 127 **Lightning Electromagnetics Volume 1 Return Stroke Modelling and Electromagnetic Radiation, 2nd Edition** V. Cooray, F. Rachidi and M. Rubinstein (Editors)
Volume 127 **Lightning Electromagnetics Volume 2 Electrical Processes and Effects 2nd Edition** V. Cooray, F. Rachidi and M. Rubinstein (Editors)
Volume 128 **Characterization of Wide Bandgap Power Semiconductor Devices** F. Wang, Z. Zhang and E.A. Jones
Volume 129 **Renewable Energy from the Oceans: From wave, tidal and gradient systems to offshore wind and solar** D. Coiro and T. Sant (Editors)
Volume 130 **Wind and Solar Based Energy Systems for Communities** R. Carriveau and D. S-K. Ting (Editors)
Volume 131 **Metaheuristic Optimization in Power Engineering** J. Radosavljević
Volume 132 **Power Line Communication Systems for Smart Grids** I.R.S Casella and A. Anpalagan
Volume 134 **Hydrogen Passivation and Laser Doping for Silicon Solar Cells** B. Hallam and C. Chan (Editors)
Volume 139 **Variability, Scalability and Stability of Microgrids** S. M. Muyeen, S. M. Islam and F. Blaabjerg (Editors)
Volume 142 **Wind Turbine System Design: Volume 1: Nacelles, drive trains and verification** J. Wenske (Editor)
Volume 143 **Medium Voltage DC System Architectures** B. Grainger and R.D. Doncker (Editors)
Volume 145 **Condition Monitoring of Rotating Electrical Machines** P. Tavner, L. Ran, C. Crabtree
Volume 146 **Energy Storage for Power Systems, 3rd Edition** A.G. Ter-Gazarian
Volume 147 **Distribution Systems Analysis and Automation 2nd Edition** J. Gers
Volume 151 **SiC Power Module Design: Performance, robustness and reliability** A. Castellazzi and A. Irace (Editors)
Volume 152 **Power Electronic Devices: Applications, failure mechanisms and reliability** F. Iannuzzo (Editor)

Volume 153 **Signal Processing for Fault Detection and Diagnosis in Electric Machines and Systems** M. Benbouzid (Editor)
Volume 155 **Energy Generation and Efficiency Technologies for Green Residential Buildings** D. Ting and R. Carriveau (Editors)
Volume 156 **Lithium-Ion Batteries Enabled by Silicon Anodes** C. Ban and K. Xu (Editors)
Volume 157 **Electrical Steels, 2 Volumes** A. Moses, K. Jenkins, Philip Anderson and H. Stanbury
Volume 158 **Advanced Dielectric Materials for Electrostatic Capacitors** Q Li (Editor)
Volume 159 **Transforming the Grid Towards Fully Renewable Energy** O. Probst, S. Castellanos and R. Palacios (Editors)
Volume 160 **Microgrids for Rural Areas: Research and case studies** R.K. Chauhan, K. Chauhan and S.N. Singh (Editors)
Volume 161 **Artificial Intelligence for Smarter Power Systems: Fuzzy logic and neural networks** M.G. Simoes
Volume 165 **Digital Protection for Power Systems 2nd Edition** Salman K. Salman
Volume 166 **Advanced Characterization of Thin Film Solar Cells** N. Haegel and M. Al-Jassim (Editors)
Volume 167 **Power Grids with Renewable Energy Storage, integration and digitalization** A. A. Sallam and B. OM P. Malik
Volume 169 **Small Wind and Hydrokinetic Turbines** P. Clausen, J. Whale and D. Wood (Editors)
Volume 170 **Reliability of Power Electronics Converters for Solar Photovoltaic Applications** F. Blaabjerg, A.l Haque, H. Wang, Z. Abdin Jaffery and Y. Yang (Editors)
Volume 171 **Utility-Scale Wind Turbines and Wind Farms** A. Vasel-Be-Hagh and D.S.-K. Ting
Volume 172 **Lighting Interaction with Power Systems, 2 Volumes** A. Piantini (Editor)
Volume 174 **Silicon Solar Cell Metallization and Module Technology** T. Dullweber (Editor)
Volume 175 **n-Type Crystalline Silicon Photovoltaics: Technology, applications and economics** D. Munoz and R. Kopecek (Editors)
Volume 178 **Integrated Motor Drives** Xu Deng and B. Mecrow (Editors)
Volume 180 **Protection of Electricity Distribution Networks, 4th Edition** J. Gers and E. Holmes
Volume 182 **Surge Protection for Low Voltage Systems** A. Rousseau (Editor)
Volume 184 **Compressed Air Energy Storage: Types, systems and applications** D. Ting and J. Stagner
Volume 186 **Synchronous Reluctance Machines: Analysis, optimization and applications** N. Bianchi, C. Babetto and G. Bacco
Volume 191 **Electric Fuses: Fundamentals and new applications 4th Edition** N. Nurse, A. Wright and P.G. Newbery
Volume 193 **Overhead Electric Power Lines: Theory and practice** S. Chattopadhyay and A. Das
Volume 194 **Offshore Wind Power Reliability, Availability and Maintenance, 2nd Edition** P. Tavner
Volume 196 **Cyber Security for Microgrids** S. Sahoo, F. Blaajberg and T. Dragicevic
Volume 198 **Battery Management Systems and Inductive Balancing** A. Van den Bossche and A. Farzan Moghaddam
Volume 199 **Model Predictive Control for Microgrids: From power electronic converters to energy management** J. Hu, J. M. Guerrero and S. Islam
Volume 204 **Electromagnetic Transients in Large HV Cable Networks: Modeling and calculations** Ametani, Xue, Ohno and Khalilnezhad
Volume 208 **Nanogrids and Picogrids and Their Integration with Electric Vehicles** S. Chattopadhyay
Volume 210 **Superconducting Magnetic Energy Storage in Power Grids** M. H. Ali
Volume 211 **Blockchain Technology for Smart Grids: Implementation, management and security** H.L. Gururaj, K.V. Ravi, F. Flammini, H. Lin, B. Goutham, K.B.R. Sunil and C. Sivapragash
Volume 212 **Battery State Estimation: Methods and models** S. Wang

Volume 215 **Industrial Demand Response: Methods, best practices, case studies, and applications** H.H. Alhelou, A. Moreno-Muñoz and P. Siano (Editors)
Volume 213 **Wide Area Monitoring of Interconnected Power Systems 2nd Edition** A. R. Messina
Volume 217 **Advances in Power System Modelling, Control and Stability Analysis 2nd Edition** F. Milano (Editor)
Volume 225 **Fusion-Fission Hybrid Nuclear Reactors: For enhanced nuclear fuel utilization and radioactive waste reduction** W.M. Stacey
Volume 228 **Digital Technologies for Solar Photovoltaic Systems: From general to rural and remote installations** S. Motahhir (Editor)
Volume 238 **AI for Status Monitoring of Utility Scale Batteries** S. Wang, K. Liu, Y. Wang, D. Stroe, C. Fernandez and J.M. Guerrero
Volume 905 **Power System Protection, 4 Volumes**

Intelligent Control of Medium and High Power Converters

Edited by
Mohamed Bendaoud, Yassine Maleh and Sanjeevikumar Padmanaban

The Institution of Engineering and Technology

Published by The Institution of Engineering and Technology, London, United Kingdom

The Institution of Engineering and Technology is registered as a Charity in England & Wales (no. 211014) and Scotland (no. SC038698).

First published 2023

The Institution of Engineering and Technology
Futures Place
Kings Way, Stevenage
Hertfordshire SG1 2UA, United Kingdom

www.theiet.org

British Library Cataloguing in Publication Data
A catalogue record for this product is available from the British Library

ISBN 978-1-83953-740-0 (hardback)
ISBN 978-1-83953-741-7 (PDF)

Typeset in India by MPS Limited

Cover Image: ghornephoto via Getty Images

Contents

About the editors

Mohamed Bendaoud is an assistant professor at the National School of Applied Sciences (ENSA) of Khouribga, Sultan Moulay Slimane University, Morocco. He is an expert on control and simulation. He has served and continues to serve on the technical program committees and as a reviewer for numerous international conferences and journals such as *Journal of The Franklin Institute, Mechatronic Systems and Control.*

Yassine Maleh is an associate professor of cybersecurity and IT governance at Sultan Moulay Slimane University, Morocco. Previous positions include lecturer at the Mohammed VI International Academy of Civil Aviation (AIAC), and the International University of Rabat. He has served on Program Committees of more than 20 conferences and events and currently as the editor for several journals, including *IEEE Access*. He received Publons Top 1% reviewer award for the years 2018 and 2019.

Sanjeevikumar Padmanaban is a full professor in electrical power engineering at the Department of Electrical Engineering, Information Technology, and Cybernetics, University of South-Eastern Norway, Norway. Previously he was with Aalborg University and Aarhus University as a faculty member – he is an editor of the *IEEE Transaction on Industry Applications, IET Electronics Letters, IETE Journal of Research,* and the subject editor for *IET Renewable Power Generation, IET Generation, Transmission and Distribution,* and *FACETS Journal* (Canada).

Preface

The importance of power converters in power processing applications has increased in recent years with the proliferation of wind power and photovoltaic, electric vehicles, microgrids, and DC distribution systems. These advances have gone hand in hand with the development and application of control technologies and the development of high-performance digital devices for integration. Although traditional control technologies have been proven to be well-established in the power electronics industry for many years, recent contributions have improved the performance of these techniques highlighting many opportunities. Additionally, the application of advanced control technologies has proven to be effective in exploiting the additional features of power converters, improving dynamic performance, efficiency, reliability, and power density.

Recently, several intelligent control techniques have been applied. A large number of papers and reports have been published. In this regard, this edited book presents the state-of-the-art and the latest discoveries in power converters. It describes and discusses new intelligent control methods applied to power converters and it presents a comparison of control methods for different converters in order to conclude the best suited for each type of converter.

The book contains 10 chapters. The first chapter presents the state-of-the-art of DC–DC and DC–AC converters with respect to their constructions, classifications, topologies, applications, and challenges. The second chapter explains the process of designing sliding mode controllers for bidirectional converters. The sliding mode parameters were chosen using the Harris Hawks optimization algorithm. The third chapter presents a novel approximation-based extremum-seeking control (AESC) method for the maximum power point tracking (MPPT) problem for the solar photovoltaic (PV) system. The fourth chapter proposes a novel control technique to implement a discontinuous conduction mode in a synchronous boost converter.

Concerning the second type of converters studied in this book, a brief overview of various inverter topologies along with a detailed study of the control architecture of grid-connected inverters is presented in Chapter 5.

The sixth chapter proposes a sliding mode control (SMC) for voltage source inverter (VSI) to regulate the powers injected into the grid. The seventh chapter presents another type of SMC used to control a three-phase natural clamped point (NPC) inverter.

The eighth chapter presents a neuro sliding mode strategy to control grid-connected three-phase inverter. A radial basis function neural network (RBFNN) architecture is utilized to estimate the unknown time-varying disturbances.

The low switching frequency operation of multilevel converters by selectively controlling the harmonic magnitude from the output waveform has been discussed in Chapter 9. Intelligent solving techniques have been employed to obtain optimal switching angles.

The last chapter gives a comparison of control methods for different converters in order to conclude the best suited for each type of converter.

In the end, the editors would like to thank all the contributors for their excellent work and cooperation in preparing this book. They further acknowledge with gratitude the assistance of the editorial and production staff at the Institution of Engineering and Technology.

Editors
Mohamed Bendaoud
ENSA Khouribga, Sultan Moulay Slimane University
Khouribga, Morocco
Yassine Maleh
ENSA Khouribga, Sultan Moulay Slimane University
Khouribga, Morocco
Sanjeevikumar Padmanaban
University of South-Eastern Norway
Porsgrunn, Norway

Chapter 1

Power electronics converters—an overview

Alireza Rajabi[1], *Milad Ghavipanjeh Marangalu*[2], *Farzad Mohammadzadeh Shahir*[2] *and Reza Sedaghati*[3]

The importance of power converters in power processing applications has increased in recent years with the proliferation of wind power and photovoltaic, electric vehicles, microgrids, and direct current (DC) distribution systems. Power converters are devices that are used to convert electrical power from one form to another. There are many types of power converters, including DC–DC converters, alternating current (AC)–DC converters, DC–AC converters, and AC–AC converters.

This chapter presents the state-of-the-art of DC–DC and DC–AC converters with respect to their constructions, classifications, topologies, applications, and challenges.

1.1 Introduction

The field of power electronics is mainly concerned with the conversion of power from one form to another and the transition from one voltage level to another using different power electronics converters. DC–DC converters with output voltage adjustability are used in a wide range of applications as the main part of many power electronic systems such as Asymmetric Digital Subscriber Line (ADSL) modems, Light Emitting Diode (LED) products, mobile phones, car electronic devices, portable devices, and renewable sources [1]. In applications that require no electrical isolation, the non-isolated high step-up converters have drawn the attention of many researchers because of their small size and simple structure. In recent years, for achieving a wider voltage gain range, isolated converters have been more widely implemented in DC–DC converters [2]. An isolated power supply provides safety by creating a strong barrier between the dangerous voltages and the input side. The drawback of isolated power supplies is lower efficiency and larger dimensions compared to non-isolated power supplies. This results from requiring

[1]Department of Electrical and Electronics Engineering, Shiraz University of Technology, Iran
[2]Faculty of Electrical and Computer Engineering, University of Tabriz, Iran
[3]Department of Electrical Engineering, Beyza Branch, Islamic Azad University, Iran

an isolation transformer. Isolated power converters are mainly used to provide safety requirements. In converters powered by high voltages (e.g., an AC–DC converter powered by the utility network), the input is isolated from the dangerous output voltage. An important advantage of using isolated converters is that their step-up or step-down boundaries can be well adjusted by changing the number of turns [3]. Both high voltage gain and electrical isolation are needed in industrial applications. Therefore, transformer-based electrically isolated converters have been extensively studied in the literature [4]. By changing the turns ratio of the transformer in isolated converters, their voltage gain can be adjusted [5]. However, leakage inductance may occur due to inappropriate coupling strength. This can lead to a high voltage spike in active switches [6]. Furthermore, the discontinuous and uneven input current may result in electromagnetic interference (EMI) in isolated converters [7]. Consequently, to properly eliminate EMI, bulky input filters are required in these converters [8]. DC–AC converters, also known as inverters, are devices that are used to convert DC power to AC power. They are commonly used in a variety of applications, including renewable energy systems, backup power systems, and portable power systems. DC–AC converters are typically classified based on the type of AC output they produce, such as sine wave, modified sine wave, or square wave. The type of AC output produced by the converter affects the quality of the power and the compatibility of the converter with different types of loads. Inverters are used in applications, including air conditioning, uninterruptible power supply (UPS), high-voltage dc power (HVDC) transmission lines, electric cars, battery storage, and solar panels [8].

This chapter provides a thorough overview of the various converters. First, the structure of direct converters is introduced and their structure is checked. Then, their use, advantages and disadvantages are analyzed. In addition, the structure of alternating current converters has been analyzed in more detail such as application, advantages, disadvantages, and structure.

The rest of the chapter is organized as follows. In Section 1.2, the topology, equivalent circuit, performance, and application of various DC–DC converters including non-isolated DC–DC converters, isolated DC–DC converters, and resonant converters are presented. The operation of three-phase inverters, classification of two-level three-phase inverters, the types of multilevel inverters (MLIs), and various novel MLI topologies are discussed in Section 1.3. Finally, the conclusion is explained in Section 1.4.

1.2 DC–DC converters

The general categorization of DC–DC converters includes non-isolated and isolated types. Isolated converters use a transformer to isolate the input side from the output based on galvanic isolation [1]. In this case, the output is not affected by the input side of the isolated converters. The polarity of the converter output can be either negative or positive depending on the configuration of the converter. The converter can function over a wide input range and provide a stable output. Some serious limitations of isolated converters are high voltage spikes on switches, thermal effect, core saturation, and leakage inductance. Moreover, these converters have large sizes that make them costly in comparison with non-isolated converters. No input/output galvanic isolation is used in non-isolated converters, so the fluctuations on the input side have a direct effect on the output side. Additionally, the number of components in non-isolated converters is less than that in isolated converters [2]. However, some small problems, such as requiring extra circuitry for optimal operation, poor voltage gain, and high duty cycle ratio, have to be resolved in non-isolated converters. In Figure 1.1, the family of power converters is shown by including the conventional converters from both categories.

1.2.1 Non-isolated DC–DC converters

The advantages of non-isolated DC–DC converters are more than those of isolated converters. Despite the fact that non-isolated DC–DC converters have some small problems, such as requiring additional circuits, low voltage gain, and high duty cycle ratio, they are a better alternative to isolated converters.

1.2.1.1 Buck converters

The DC–DC buck converter steps down the output voltage level with respect to the input voltage level (Figure 1.2) [1]. Thus, this type of converter can be used to integrate a larger module voltage to smaller battery voltages or loads.

1.2.1.2 Boost converters

In some applications, the load side voltage should be greater than the output voltage. Therefore, boost converters can be implemented in maximum power point

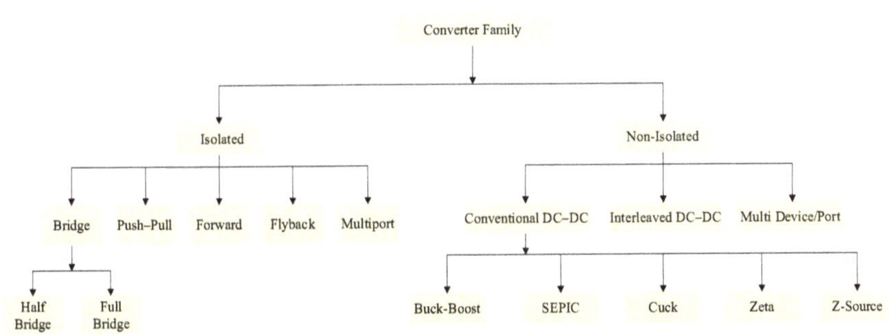

Figure 1.1 Classification of power converters

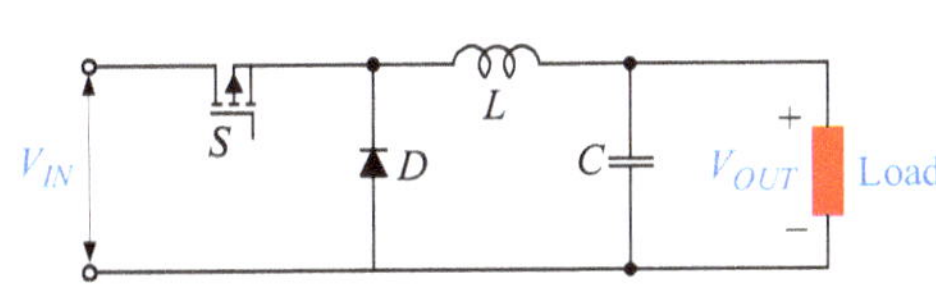

Figure 1.2 Buck converter topology

tracking (MPPT) converters. Many studies have focused on boost converters, and these converters have been modified for achieving high performance (Figure 1.3) [2]. In this topology, the inductor is series with input voltage and it can reduce the current stress of the switch.

1.2.1.3 Buck–boost converters

A buck–boost converter is shown in Figure 1.4. This converter is realized by integrating a basic buck converter with a boost converter. The final design can be used in numerous applications, including standalone or grid-connected photovoltaic (PV) systems and motor drives.

Under the continuous conduction mode (CCM) of buck–boost converters, current ripples are small. The voltage and current stress on the operating elements of buck–boost converters that use two switches is smaller compared to single-switch buck–boost converters. Moreover, the measurement results indicate the higher efficiency of buck–boost converters under heavy load operation [5].

1.2.1.4 Single-ended primary inductance converters

Figure 1.5 shows a single-ended primary inductance converter (SEPIC). In the switching process of this converter, the OFF time has to be shorter than the ON time, so a higher output voltage is achieved since the time required to charge the inductor is longer. Otherwise, the converter cannot supply the needed output since the capacitor cannot be completely charged.

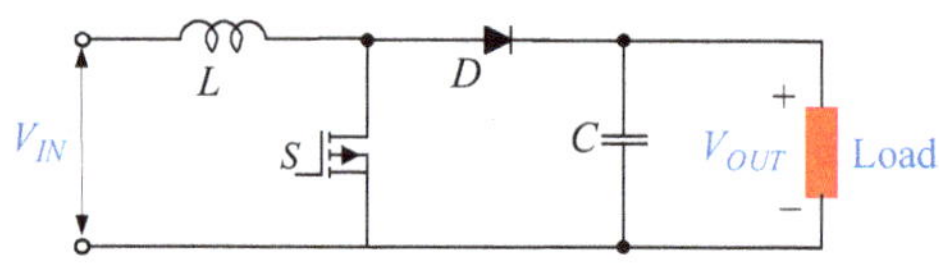

Figure 1.3 Boost converter topology

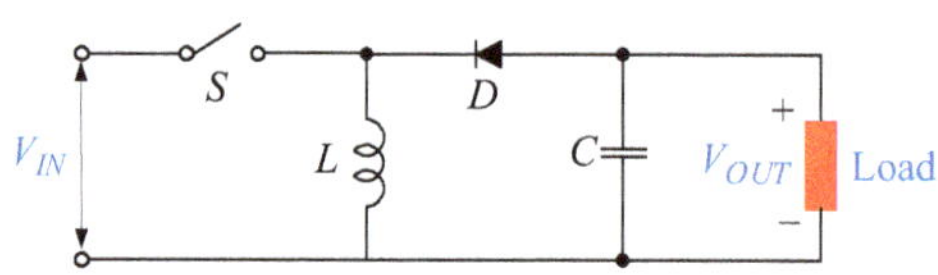

Figure 1.4 Buck–boost converter topology

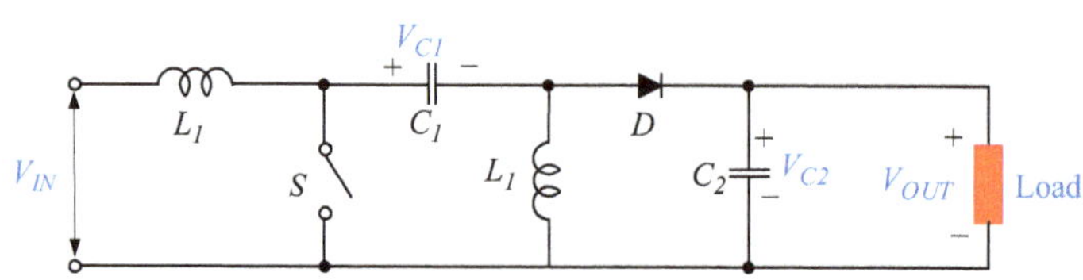

Figure 1.5 SEPIC converter topology

During the design process of converters, some constraints have to be taken into account. When a high-frequency transformer is used in a conventional SEPIC, the reduction in the output voltage ripple is feasible. This structure has some advantages such as continuity of output current besides lower output ripple and switching stress [6]. Some harmonics are induced in AC–DC conversion. This results in AC ripple current and a reduction in power factor. When SEPIC is operated in boundary conduction mode, the power factor of AC lines is corrected. SEPIC converters are widely used in many PV system applications, such as in improving the power factor in AC lines and regulating the flickering DC voltage. Moreover, SEPIC converters have a non-inverting output, so these converters have drawn more attention than buck–boost converters and are better for high power applications.

1.2.1.5 Cuk converters

A Cuk converter topology is shown in Figure 1.6. The Cuk converter is similar to a buck–boost converter. However, the inductor in buck–boost converter is replaced with a capacitor to transfer power and store energy. In a Cuk converter, the polarities of input and output voltages are reversed. The converter output can be inverted by appropriate connections. In this case, the output exhibits no ripple, which is useful for many applications [7]. Cuk converters are used in power factor correction (PFC) and voltage regulation in many applications. The periods of ON time and OFF time of the control switch define voltage boosting. The switch capacitors discharge within the ON time period and inductors store energy. Within the OFF time period of a switch, the diode operates in conduction mode and allows the flow of current. In Cuk converters, capacitors act as energy storage components, while in other converters, inductors are used as energy storage components.

1.2.1.6 Z-Source converters

Another efficient converter topology is Z-source, which can step up and step down the output voltage. As shown in Figure 1.7, in the structure of Z-source converters, a specific inductance–capacitance (LC) impedance scheme is used to connect the main circuit of the converter to the power source. For high and medium power applications, Z-source converters are more preferable. The output ripple noise of these converters is low while their duty cycle is lower than 0.5. In comparison with conventional boost converters, Z-source converters can more effectively boost the voltage at the same duty cycle. In addition to acceptable efficiency, the cost and dimensions of these converters are more efficient than other converters. A modified Z-source topology with low switching stress and reduced size is integrated to

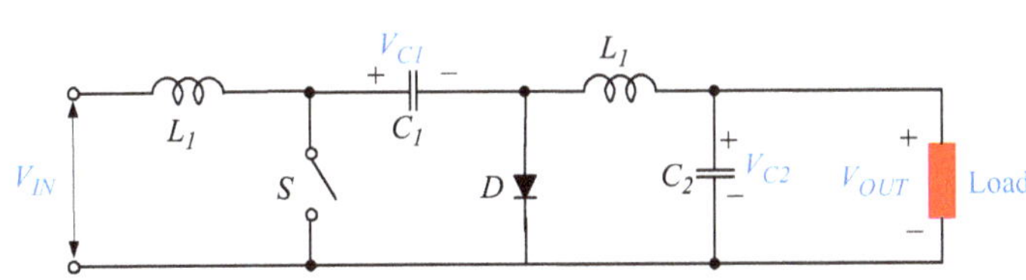

Figure 1.6 Cuk converter topology

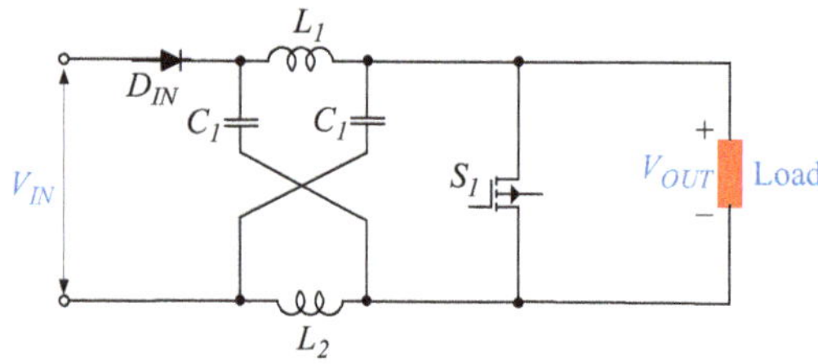

Figure 1.7 Z-source converter topology

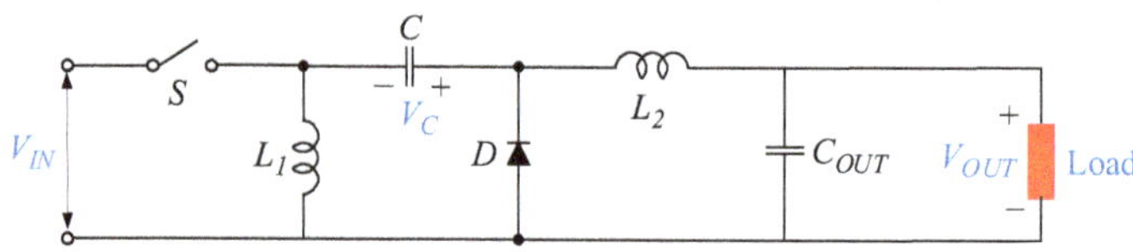

Figure 1.8 Zeta converter topology

generate PV power using the same input and output ground. A Z-source topology is proposed in [8] to keep the input and output fuel cell variables constant as well as to obtain a constant DC link voltage.

1.2.1.7 Zeta converters

Figure 1.8 shows a zeta converter. This converter provides a non-inverted output voltage while stepping down or up the input voltage, in a similar manner as in SEPIC. The main characteristic of zeta converters that are implemented in PV systems is their capability to track MPP on the whole range of PV curves [9]. Zeta converters are also famous for their power optimization capability. The number of components in these converters is relatively small. Zeta converters have special properties such as continuous output current, regulated output voltage, and non-inverting output. Moreover, easier compensation and lower output ripple are other advantages of these converters. Reference [10] investigates a zeta converter that is integrated with a PV system to energize brushless DC (BLDC) motors for water pumping. This design allows maximum power utilization and low power loss of PV cells while no extra control circuit is required. In [11], a constant output is achieved for zeta converters under load variations in wind turbine applications.

1.2.1.8 Hybrid non-isolated DC–DC converters

Conventional non-isolated converters suffer from issues that reduce the service life of the equipment integrated into the system and affect the efficiency of the converter. Therefore, converter topologies can be hybridized to achieve new and more efficient converter topologies. This is done by considering the limitations and properties of converters topologies that are being hybridized. Non-isolated converters have some issues, such as high duty cycle ratio, higher switching stress, discontinuous output/input current, and high output/input ripple. Moreover, Table 1.1 presents a list of the

Table 1.1 Characteristics of non-isolated DC–DC converters

Converter topology	**Specifications**	**Advantages**	**Drawbacks**
Buck–boost converter	• Low complexity • Easily controlled • Small dimensions • Cost efficiency	• Appropriate for low power applications • Compatible to work with high switching frequencies • Non-linear relationship between output voltage and duty cycle	• Voltage imbalance for multi-input and multi-output applications • Large output ripple • High duty cycle ratios required for high voltage gain • Discontinuous output current
Single-ended primary Inductor converter (SEPIC)	• Low complexity • Easy to control • Small dimensions • Cost efficiency	• Non-inverting output • Used to correct power factor in AC lines • Non-linear relationship between output voltage and duty cycle	• Difficulty in controlling duty cycles for multi-input and multi-output configuration • Low voltage gain • Duty cycle imbalance to effectively operate and the ON time has to be greater than OFF time to obtain higher output voltage
Cuk converter	• Low complexity • Easy to control • Small dimensions • Cost efficiency	• Appropriate for low power applications • Using capacitor for power transfer and energy storage • Non-linear relationship between output voltage and duty cycle	• Negative polarity of output compared to input • Efficiency is reduced in multiple output network • Complex compensation circuitry is required for proper operation of the converter • Discontinuous output current that cannot be controlled because of the resonance of L-C pair, which results in excessive voltage on capacitor and probable damage to the circuit
Z-Source converter	• Moderate complexity • Easily controllable • Small dimensions • Cost efficiency	• Non-inverting output • Applicability to power conversion • Non-linear relationship between output voltage and duty cycle	• Unidirectional power flow • Input current discontinuity • Large reverse recovery issue • Difficulty in synchronizing multiple output networks • No soft-start capability

(Continues)

Table 1.1 (Continued)

Converter topology	Specifications	Advantages	Drawbacks
Zeta converter	• Moderate complexity • Easily controllable • Small dimensions • Cost efficiency	• Applicability to medium- and high-power systems • Non-inverting output • Applicability to power conversion • Non-linear relationship between output voltage and duty cycle	• Unidirectional power flow • Voltage imbalance for multiple output • Mostly requiring compensation circuit
High step-up DC–DC converter	• Moderate complexity • Requiring precise control • Small dimensions • Moderate • cost	• Non-inverting output • Applicability to renewable energy applications • Requiring small capacitive filter values • Common input and output ground • More operation modes • Low input ripple and soft-switching	• Applicability to multi-input single-output structure • Requiring effective control techniques to control double duty cycle • No more than 50% duty cycle of each switch under interleaved control with 180 degree phase shift • Input conduction loss because of coupled inductors
Input-parallel output-series high-gain DC–DC converter	• Medium complexity • Requiring precise control • Medium dimensions • Moderate cost	• Non-inverting output • Applicability to renewable energy systems • Input current sharing • More operation modes • Low output ripple and less reverse recovery • period	• Unidirectional power flow • Requiring effective control techniques to control double duty cycle • Transients because of diode reverse recovery issue and capacitor charging • Switching duty cycle transition
Double-duty-triple-mode high-gain transformer-less DC–DC converter	• Medium complexity	• Non-inverting output	• Unidirectional power flow

(Continues)

Table 1.1 (Continued)

Converter topology	Specifications	Advantages	Drawbacks
	• Requiring precise control • Medium dimensions • Moderate cost	• Applicability to renewable energy systems • Duty cycle with wide range • More operation modes • High output gain with no need for complex technique	• Requiring efficient control techniques for controlling two duty cycles • Transients because of diode reverse recovery issue and capacitor charging • Voltage fluctuation in switching modes
Three-state switching high-gain hybrid boost converter	• Moderate complexity • Requiring precise control • Moderate dimensions • Medium cost	• Non-inverting output • Applicability to renewable energy systems • Lower duty cycle ratios • More operation modes • High output gain with • Voltage lift technique	• Unidirectional power flow • Requiring efficient control techniques for controlling two duty cycles • Two duty cycles • Transients because of diode reverse recovery issue and capacitor charging • Switching duty cycle transition

disadvantages and advantages of non-isolated converters. Figure 1.9 shows the state-of-the-art hybrid non-isolated DC–DC converters. Figure 1.9(a) shows a high step-up hybrid DC–DC converter topology, which can both step up and step down the voltage [12]. In addition to using a common output/input ground, which results in grid-connected transformer-less topology, the high step-up hybrid DC–DC converter topology can be effectively used for PV generation applications. This topology can also perform soft switching that allows zero current switchings (ZCS) and zero voltage switching (ZVS) of passive and active elements. The circuit design has two series coupled inductors at the output for achieving the highest possible voltage gain. In addition, in intermittent conditions, the circuit operates at a minimum duty cycle while delivering continuous output and high gain. Figure 1.9(b) shows a high-gain output-series input-parallel DC–DC converter. This hybrid topology integrates parallel input and series boost converter consisting of a voltage multiplier and two coupled inductors [13]. On the input side, two inductors are connected in parallel and share the input voltage and input current ripple. Interleaved series capacitors are integrated on the output side. This converter can be employed in residential or industrial applications. Furthermore, this topology has high voltage gain, low output ripple, and nominal switching stress. Figure 1.9(c) illustrates a double-duty-triple-mode transformer-less

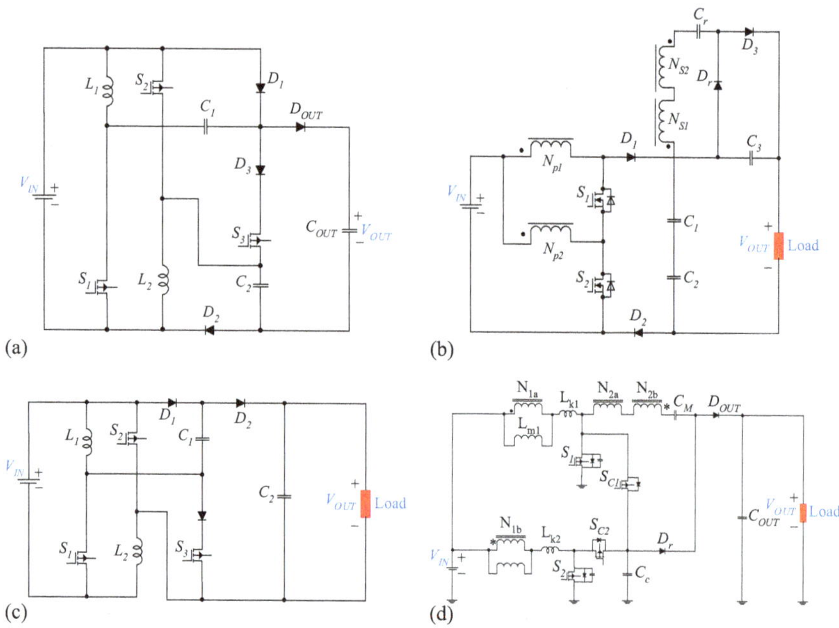

Figure 1.9 Hybrid non-isolated converters: (a) hybrid high-gain three-state switching boost converter [15], (b) double-duty-triple-mode high-gain transformer-less DC–DC converter [14], (c) high-gain output-series input-parallel DC–DC converter [13], and (d) high step-up DC–DC converter [12]

high-gain DC–DC converter [14]. The number of components is relatively small in this converter while its efficiency is higher than that of conventional non-isolated converters. This topology can operate at two duty cycles, which can mitigate the adverse effects of high duty cycle ratios. Moreover, three operation modes are considered for this topology, and it can perfectly exploit the power electronic components so that a more stable output and higher voltage gain are achieved. Therefore, this topology requires no complex circuit with voltage lift, coupled transformers, or voltage multiplier. The applicability of this topology for DC microgrids is confirmed experimentally. As a result, this topology can be implemented in renewable energy systems for proper operation and service life improvement of PV cells. Reference [15] proposes a non-isolated converter called three-state switching high-gain hybrid boost converter [see Figure 1.9(d)]. This topology has a similarity to hybrid converters mentioned earlier with the only difference that this converter uses an extra voltage lift circuit to achieve greater voltage gain. Additionally, this converter has a small number of components and a high voltage gain with an optimal switching duty cycle ratio. The use of three operation modes and two duty cycle ratios provides excellent voltage boosting performance. Furthermore, this converter can be used in high-efficiency voltage boosting in renewable energy applications including fuel cells and PV systems.

1.2.2 Isolated DC–DC converters

Isolated DC–DC converters are crucial in applications where safety is a critical aspect. This is essential in preventing the impact of dangerously high output voltages on the input side. However, non-isolated converters can be employed in applications like the power supply of an X-ray system. Isolated DC–DC converters have key role when regulations or safety considerations are necessary. However, transformer-based isolation leads to some design challenges in terms of assembly, performance variability, size, and cost. Isolated DC–DC converter topologies can be divided in four main groups: forward, flyback, full-bridge, half-bridge, and push-pull. These types of converters are introduced as step-down converters that have been used in different applications, such as fast charging systems or power supplies. The importance of isolation in DC–DC converters is described in detail in the following.

1.2.2.1 Flyback converters

Figure 1.10 illustrates a flyback converter, which is largely used in ultra-low-power PV systems. To achieve a higher conversion gain while transformers are employed, the best option is to use a flyback converter. Transformers need a large airgap for energy storage in high-power applications. The magnetic inductance is smaller when the airgap is large, and a flyback converter delivers a larger leakage flux and transfers energy with very low efficiency. Cuk converters are employed in high-power applications. However, these converters have some disadvantages, including higher current flow in the diode/power switch (at the output) and output polarity inversion. Reference [16] investigated the unique features of flyback converters, such as swift dynamic response and low system complexity. Moreover, ZVS operation can enhance the efficiency of these converters [16]. It is possible to achieve soft switching by using a clamp circuit plus a flyback converter that is based on resonance.

A flyback converter stores energy during the switch's ON state and transfers the energy to the load during the OFF state. Flyback converters are noncomplex switch-mode power supplies, which can generate DC outputs from both DC and AC inputs. During the off-state of the main switch, flybacks transfer power from the input to the output. Flyback converters have some drawbacks, such as pulsating

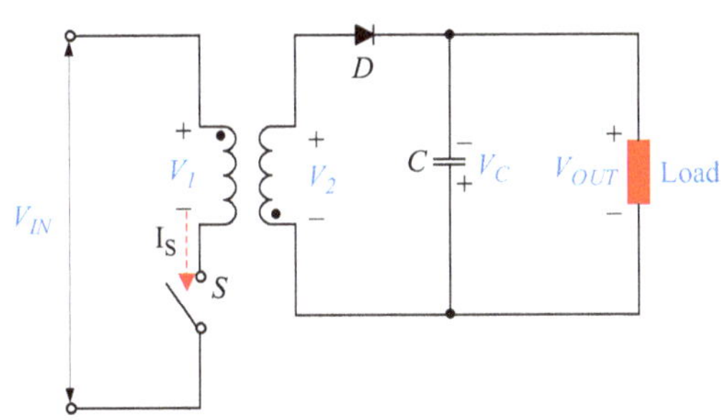

Figure 1.10 Flyback converter topology

source current, requiring extra snubber circuits for overcoming the inductor leakage current, high RMS current rating of output capacitors, and poor efficiency. In switched-mode power supplies, flyback transformers (or equivalently, line output transformers) store energy from the input voltage.

1.2.2.2 Bridge converters

Generally, DC–DC bridge converters include two or four active switches in the bridge structure across power transformers in high-frequency applications.

Half-bridge converters

In the same way as forward and flyback converters operate, a half-bridge topology can provide an output side lower or higher than the input voltage and use a transformer for achieving electrical isolation. This structure is also referred to as "half bridge." A half-bridge converter operates as an electronic toggle switch, while it cannot reverse the voltage polarity on the load. Half-bridge converters are implemented in switched-mode power supplies based on switching amplifiers and synchronous rectifiers. A three-port half-bridge topology is shown in Figure 1.11 for renewable energy applications. The main circuit of a half-bridge converter acts as a buck converter with synchronous rectification, which allows the flow of DC bias current through a high-frequency transformer. In half-bridge converters, synchronous regulations and post-regulation are considered. Moreover, these converters can be independently regulated on three ports. Accordingly, the advantages of half-bridge converters are single-stage power conversion and simple control [17]. Three switches are used in active-clamped half-bridge DC converters, which show input current continuity with low ripple and wide-range ZVS.

Full-bridge converter

These converters are usually used to provide isolation in medium to high power systems and/or to step down high DC bus voltage. Full-bridge converters are used in renewable energy systems, battery charging systems, telecom rectifiers, and server power supplies. They are primarily designed to convert AC input from mains

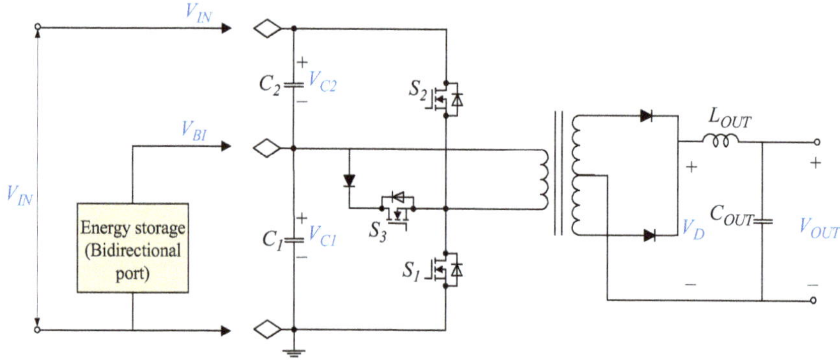

Figure 1.11 Three port half-bridge converter topology

power to a DC output, i.e., usable device power. Bridge rectifiers are electronic components widely used in many different types of circuits and power supplies (PSUs) and are found in all workplace appliances and electrical products. Figure 1.12 shows a full-bridge converter topology, which is normally implemented in renewable energy units as the interface for the load, storage device, and renewable source. Two buck–boost converters are integrated with full-bridge topology to realize a three-port full-bridge converter. This structure easily enables ZVS of the existing switches in a single conversion stage. A full-bridge asymmetrical pulse-width-modulated converter with high efficiency is proposed to be used in renewable power systems. This design, which is based on an asymmetric control and full-bride topology, can provide zero-voltage switching. This is feasible by turning on the switches that minimize the loss due to the circulating current [18].

Dual active bridge converter

Figure 1.13 shows an isolated bidirectional DC–DC converter (IBDC) with dual active bridge (DAB). Many researchers have focused on this design, which possesses special features such as inherent soft switching, bidirectional power flow, high power density, galvanic isolation, and high conversion efficiency. These

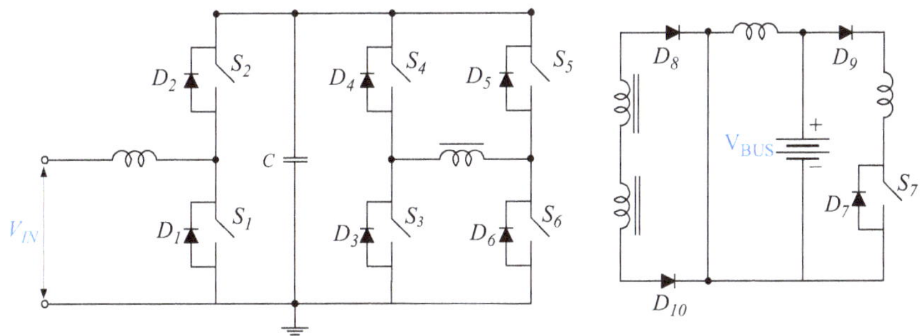

Figure 1.12 Full-bridge converter topology for solar applications

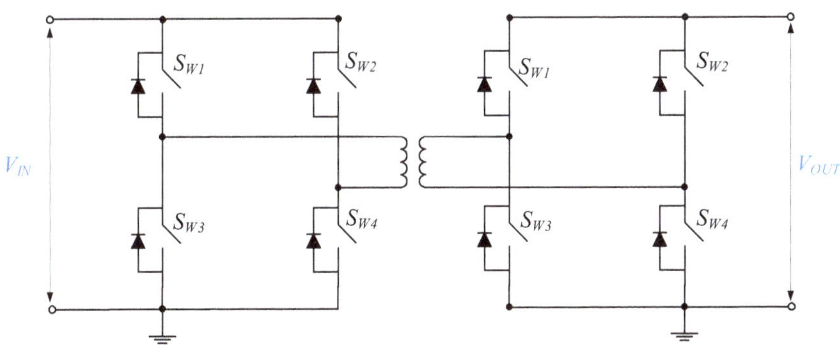

Figure 1.13 DAB converter topology

properties allow DAB–IBDCs to be implemented in standalone hybrid systems. To isolate two full bridges, a high-frequency power transformer is considered. The leakage inductor can store energy. The primary and secondary bridges are connected respectively to high voltage-side DC sources and low voltage-side energy storage devices or loads. The phase of the square waves between the two bridges can be easily shifted to allow the power to flow bidirectionally. The voltage difference at the energy storage component has to be exploited for power conversion [19].

1.2.2.3 Push–pull converter

Figure 1.14 shows another DC–DC converter model called a push–pull topology. A push–pull topology is also known as a switching converter that uses transformers to transform the supplied DC power. This topology is fundamentally a forward converter that operates using a center-tapped primary winding. Therefore, the core of the transformer is used effectively in comparison with other types of converters, such as flyback or forward converters. Additionally, copper losses increase since only half of the winding copper is utilized each time. Compared to forward converters, push–pull converters require smaller filters at a certain power level.

Push–pull converters have special features, which can be summarized as follows. The primary winding uses the current flowed through a pair of transistors on the input lines in the push–pull circuit. Simultaneous switching-on-and-off of these transistors draws current on some occasions. Consequently, under switching operation, the current is broken from the line over the half-cycle pair. The push–pull input current is stable. Moreover, this converter operates with low input noise and shows higher efficiency in high-power applications [20].

In push–pull converters, transistor pairs in a symmetrical push–pull circuit supply the current to the transformer primary from the input line. The alternately switching on and off of the transistors leads to a periodical current reversal in the transformer. As a result, over both switching halves, the current is supplied by the line. A push–pull converter makes a low level of sound on the input line and has a steady input current. For achieving galvanic isolation of the modules and utility and greater step-up ratios, the HF transformer isolation technique is used in the converter.

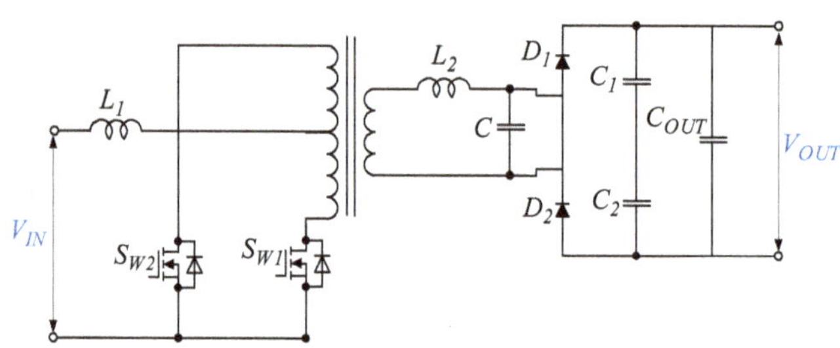

Figure 1.14 Push–pull converter topology

1.2.3 Resonant converters

Resonant power converters (RPCs) have six stages. The input source, which may be a current or voltage source, is in the first stage. A full-bridge or half-bridge inverter is used in the second stage. Moreover, a resonant tank circuit of the *N*th order ($N = 0,1,2,\ldots, N$), which is consisted of *N* capacitors and inductors, is supplied by the output of the second stage. In the subsequent stage, a stage-isolation transformer is used. Moreover, this transformer can step down and step up the voltage. The next stage consists of a half-bridge or full-bridge rectifier, which is supplied by the isolation stage's output. Then, the filter stage with a high pass filter or low pass filter is located next to the bridge rectifier stage.

1.2.3.1 Series resonant converters

Figure 1.15(a) shows a series resonant converter (SRC). This type of converter has a rectifier, isolation transformer, two-element resonant network, and a half-bridge

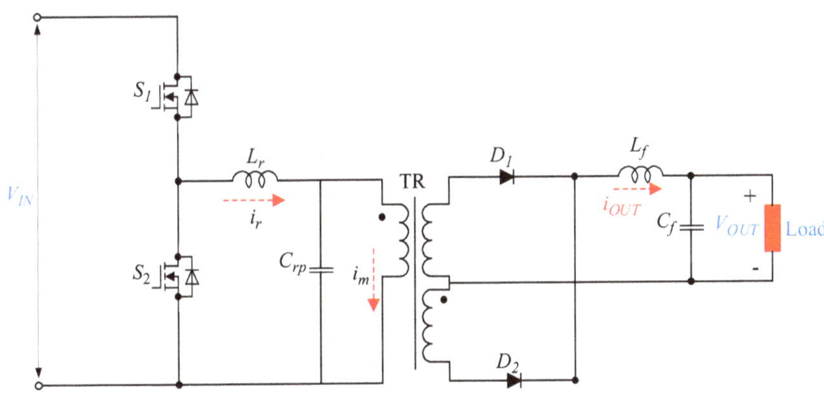

(a)

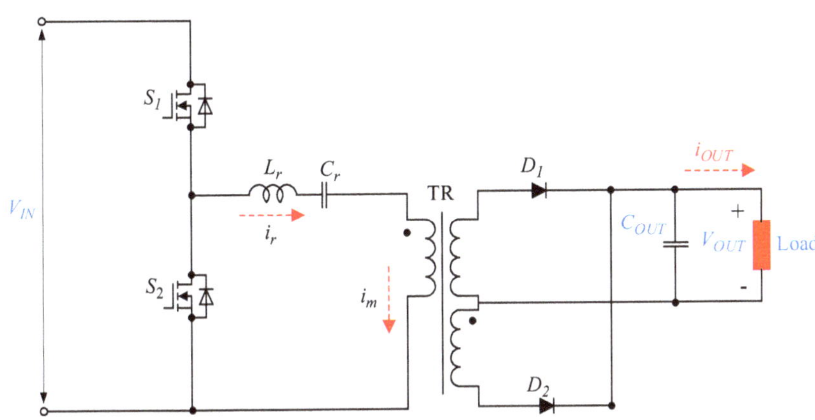

(b)

Figure 1.15 (a) Half-bridge PRC and (b) half-bridge SRC

inverter. Two switches are included in the half-bridge inverter with parasitic capacitors and a body diode. The resonant capacitor, Cr, and resonant inductor, Lr, are included in the resonant tank circuit. These resonant elements are connected in series to the transformer winding, and this is why this converter is named the "series resonant converter." Additionally, the load is connected in series to the resonant tank circuit at the rectifier output. In this design, the load and tank circuit have the role of a voltage divider. By tuning the driving voltage frequency, the resonant tank impedance can be varied. The SRC operates as a voltage divider to divide the input voltage between the resonant tank impedance and load. Consequently, the SRC always has a DC gain that is smaller than unity. The tank circuit impedance is small at resonance, so the resonant converter's output voltage is equal to the input voltage. Moreover, the SRC gain is maximum at resonance. The capacitor on the primary side of the transformer in an SRC converter blocks the primary current dc component. Hence, under no-load conditions, the voltage gain of SRC shows a low selectivity. Accordingly, the curve of the voltage gain is a horizontal line, so SRC converters have the drawback of lack of operability under no-load conditions. Another drawback of SRCs is that the high ripple currents have to be handled by an output filter. As a result, SRCs are not an appropriate option for high-current low-voltage applications [21].

1.2.3.2 Parallel resonant converters

Figure 1.15(b) shows a parallel-resonant converter (PRC) in which one or both reactive elements are parallel to the load [21]. A capacitor and an inductor are used respectively on the primary and secondary sides of the transformer for impedance matching. A PRC can be used for output voltage regulation under no-load conditions, so this limitation of SRCs is addressed. However, the drawback of PRCs is that with increasing input voltage, the circulating current magnitude increases. The circulating current magnitude in PRCs is higher than that in SRCs.

1.2.3.3 Series–parallel resonant converters

Figure 1.16 shows a series–parallel resonant converter (SPRC), which has three reactive components. An SRC is combined with a PRC to realize a hybrid resonant

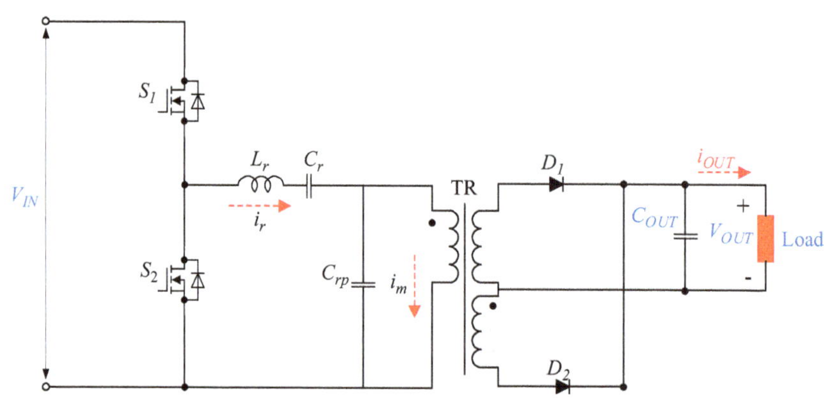

Figure 1.16 Half-bridge series–parallel converter

tank of SPRC. In a similar way to PRCs, an output filter inductor is implemented on the secondary side for impedance matching. In this configuration, the limitations of SRC and PRC, such as circulating current flow and no-load regulation, are eliminated. To this end, resonant components have to be properly designed and selected. In this SPRC, the output voltage under no-load conditions can be regulated only for very small values of Cp. For very small values of Cp, the responses of SPRC and SRC are similar [21].

1.2.3.4 LLC resonant converters

An LLC resonant converter has the advantages of both PRC and SRC. By using an LLC resonant converter, ZVS turn-on and ZCS turn-off operations can be achieved. Moreover, by operating at higher frequencies, a greater power density can be achieved, which leads to smaller transformer dimensions. Galvanic isolation is also provided by the use of the transformer. Resonant converters generate wide output voltage ranges. The resonant components impedance changes by changing the switching frequency. This, therefore, changes the converter gain. No spike is observed in the waveforms of current and voltage of the diode rectifier, so it contains small amounts of harmonic pollution and EMI. A half-bridge LLC converter is shown in Figure 1.17. This converter has five parts as follows: an output capacitor, a center-tapped rectifier circuit, a transformer with *n*:1 turns ratio, a resonant tank, and a half-bridge inverter. The LLC converter supplies batteries as resistive loads. The SRC and LLC resonant converters are different as a result of the existence of the magnetizing inductance Lm. The LLC converter includes Lm, Cr, and Lr, which denote magnetizing inductance, resonant capacitor, and resonant inductor, respectively. A half-bridge inverter is an input to the tank circuit. A capacitor in series with the power line can automatically balance the flux. This converter has two resonant frequencies and works in variable frequency modes. The first resonant frequency of the LLC converter is related to Lr and Cr, while the second resonant frequency is related to Cr, Lr, and Lm [22].

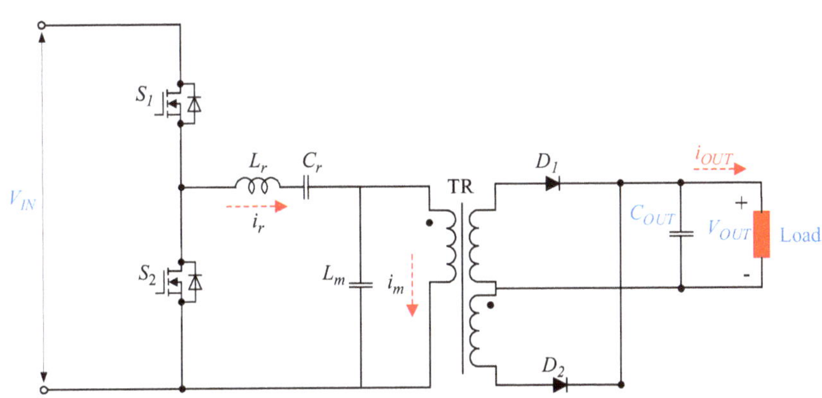

Figure 1.17 Half-bridge LLC converter

1.3 DC–AC converters

1.3.1 Two-level single-phase and three-phase inverters

DC–AC converters have been applied in grid-connected applications, electric vehicles (EV), and renewable power generation systems such as photovoltaic (PV) [23]. Inverters have been classified into square-wave topologies, sinusoidal two-level pulse width modulation (PWM)-based structures, and multilevel inverter topologies. The power electronics converters should have high-quality reliability, efficacy, and performance to create a sinusoidal current for injecting good quality power into the grid in the grid-tied PV-based systems. The electrical circuit of a two-level structure is illustrated in Figure 1.18. Regarding Figure 1.18, it is obvious that by turning on the switches S_1 and S_2 the output voltage of the inverter is equal to $+0.5V_{IN}$ and $-0.5V_{IN}$, respectively. Considering Figure 1.19, it can be understood that the two-level topologies can generate a square wave two-level output voltage waveform without using the pulse width modulation (PWM) strategy. The mentioned conventional topology operates at a high switching frequency and is suitable for high power applications. However, the mentioned structure has high power switching losses and low overall efficiency. The other disadvantages of the two-level conventional topology are the high value of total harmonic distortion (THD), and high voltage stress on used switches. Due to the mentioned disadvantages, it is not appropriate for high-voltage grid-tied power generator systems. In order to overcome the mentioned problems in the two-level conventional inverters, different topologies of multilevel inverters have emerged. Multilevel inverters have many benefits such as high switching frequency, improved power quality, lower THD, lower power losses, and higher overall efficiency. Meanwhile, in order to implement the multilevel inverter, a separate gate driver is needed for

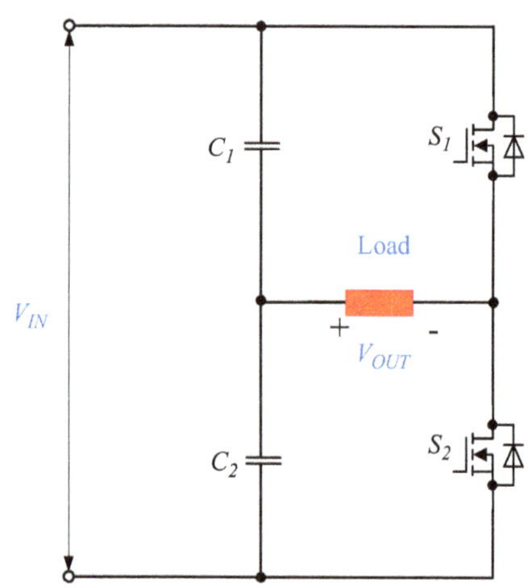

Figure 1.18 A circuit of two-level inverter

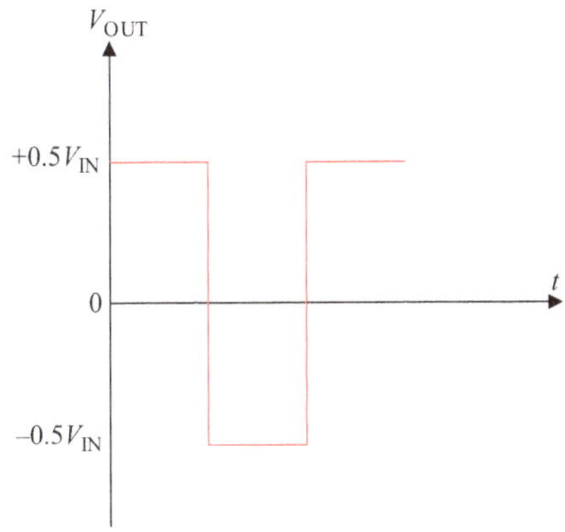

Figure 1.19 Two-level output voltage waveform

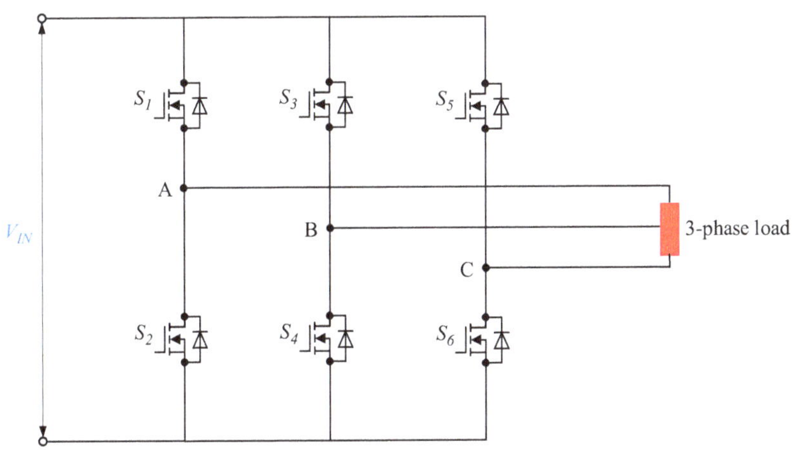

Figure 1.20 Three-phase H6 inverter

each switch, which makes the system complicated and counter-intuitive. So, it is necessary to reduce the count of power switches in the multilevel inverters.

1.3.2 Classification of two-level three-phase inverters

The three-phase two-level topology is indicated in Figure 1.20. Compared with current source inverters, the voltage source topologies have some benefits such easy control system, low cost, and being a mature technology. Therefore, these kinds of topologies are very useful in the PV-based grid-tied applications [24]. Meanwhile, the conventional three-phase two-level H6 structure has not a voltage boosting feature. Therefore, an extra dc–dc boost stage is applied to the mentioned topology to boost the input low dc voltage [25]. Furthermore, in order to overcome

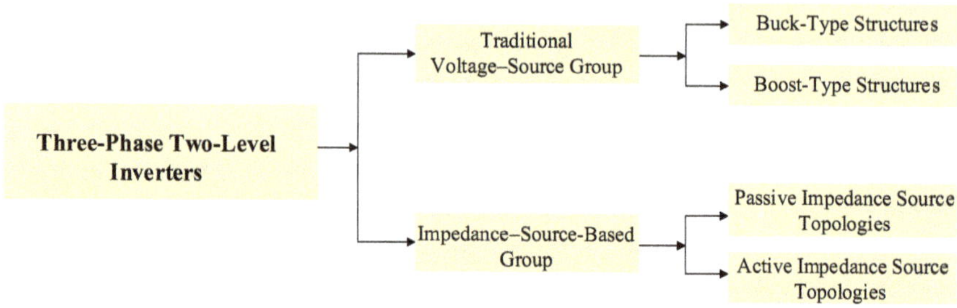

Figure 1.21 Three-phase two-level structures

to short-circuit problem of the DC-link bus, a deadtime must be applied in the H-bridge switches. This issue increases the value of THD of the output voltage waveform [26]. Nowadays, single-stage impedance source structures are proposed with enhanced reliability and buck–boost feature [27]. The major challenge in transformer-less grid-tied PV inverters is the generating of common mode voltage (CMV) and leakage current [28]. To mitigate the leakage current in transformer-less inverters, the CMV is fixed to a constant value or limited. Both the PWM control strategies and system structure reconfiguration are presented to mitigate the CMV in the conventional three-phase two-level H6 topologies [24,29]. Further, in order to reduce the CMV in the impedance source structures some solutions have been also presented. In addition, regarding the count of active power switches in the impedance source topologies, they can be categorized into passive and active types [28,30]. Therefore, the three-phase two-level topologies with limitation of leakage current are classified into two groups: (a) impedance source-based topology and (b) conventional voltage source topology. Figure 1.21 summarizes the several kinds of three-phase two-level structures.

1.3.3 Multilevel inverters

This section investigates three types of multilevel inverters (MLIs): (1) diode-clamped MLIs, (2) flying capacitor MLIs, and (3) cascaded H-bridge MLIs.

1.3.3.1 Diode-clamped MLIs

In this inverter, diodes are mainly used to reduce the voltage stress of power devices. The voltages of switches and capacitors are determined as V_{dc}. It should be noted that several elements are required in n-level inverters, including $(n-1)(n-2)$ diodes, $2(n-1)$ switching devices, and $(n-1)$ voltage sources.

Five-level diode-clamped MLI

The following specifications are considered for five-level diode-clamped MLIs:

$n = 5$;

Therefore,

Number of switches $= 2(n-1) = 8$;

Number of diodes $= (n-1)(n-2) = 12$;

Number of capacitors $= (n-1) = 4$.

Figure 1.22 illustrates a five-level diode-clamped MLI, and Table 1.2 gives the switching states. For instance, switches S1 to S4 should be simultaneously operated to obtain an output of $V_{dc}/2$. It is clear that these four switches have to be operated for each voltage level. From Table 1.2, the maximum voltage at the output is

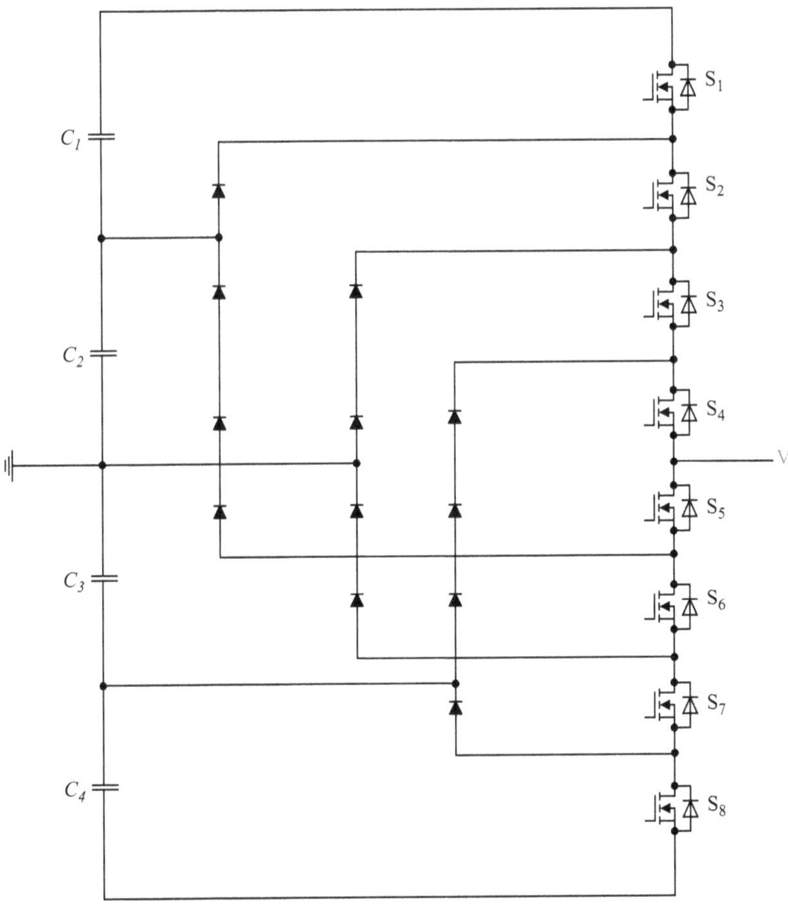

Figure 1.22 One phase of a diode-clamped inverter

Table 1.2 Diode-clamped MLI switching states

V_{out}	S_1	S_2	S_3	S_4	S_5	S_6	S_7	S_8
$V_{dc}/2$	1	1	1	1	0	0	0	0
$V_{dc}/4$	0	1	1	1	0	0	0	0
0	0	0	1	1	1	1	0	0
$-V_{dc}/4$	0	0	0	1	1	1	1	0
$-V_{dc}/2$	0	0	0	0	1	1	1	1

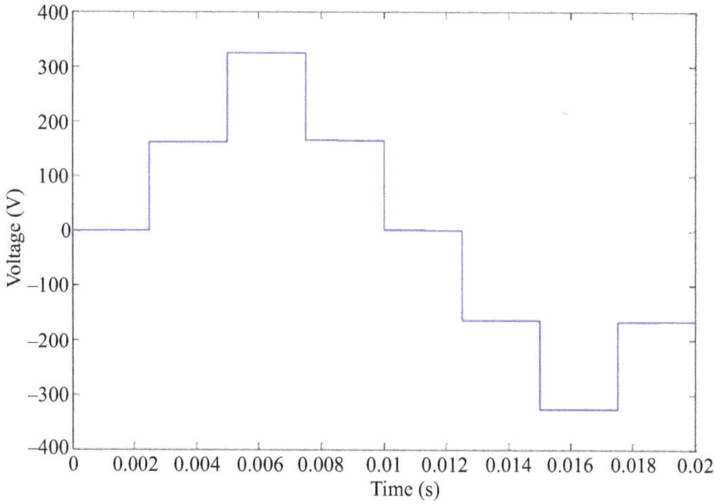

Figure 1.23 Output voltage of a five-level MLI

determined to be half of the DC source voltage. However, this is considered to be the drawback of diode-clamped MLIs. This is a problem that can be solved by cascading two-diode-clamped MLIs or using a two-time voltage source. Figure 1.23 illustrates the output voltage of a five-level diode-clamped MLI. From Figure 1.23, an equal voltage has to be considered for all voltage levels. The results of computing switching angles show the lowest THD for output voltage [31].

1.3.3.2 Flying capacitor MLIs

The purpose of using flying capacitor MLIs is to limit the voltage of power devices. The configuration of these inverters is similar to that of diode-clamped MLIs. However, capacitors are employed for input DC voltage division. The voltages of capacitors and switches are determined to be V_{dc}.

Five-level flying capacitor MLIs

The following values are considered for a five-level flying capacitor MLI:

n=5;

Therefore,

Number of switches=8;

Number of capacitors= 10.

A five-level flying capacitor MLI is shown in Figure 1.24. The switching states of this inverter and a diode-clamped MLI are similar, so four switches have to be implemented for each output voltage level. The switching states of a five-level flying capacitor-clamped MLI are given in Table 1.3.

It should be noted that the lowest amount of THD is obtained from computing the switching angles that are similar to the switching angles of the diode-clamped MLI. This technique is the same as the approach utilized in Ref. [32] for diode-clamped inverters.

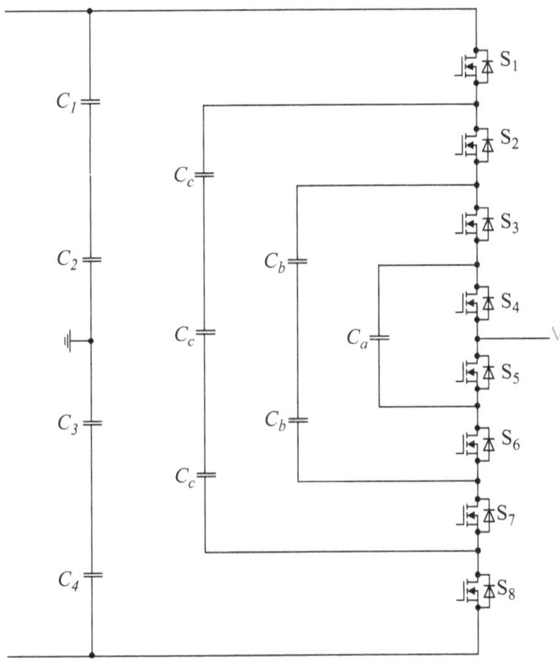

Figure 1.24 One phase of a five-level flying capacitor MLI

Table 1.3 The switch pattern of a capacitor-clamped MLI

V_{out}	S_1	S_2	S_3	S_4	S_5	S_6	S_7	S_8
$V_{dc}/2$	1	1	1	1	0	0	0	0
$V_{dc}/4$	1	1	1	0	1	0	0	0
0	1	1	0	0	1	1	0	0
$-V_{dc}/4$	1	0	0	0	1	1	1	0
$-V_{dc}/2$	0	0	0	0	1	1	1	1

1.3.3.3 Cascaded H-bridge MLIs

The purpose of applying these MLIs is to connect H-bridge inverters in series to obtain a sinusoidal output for voltage, which can be obtained by summing the voltage generated by each cell. Voltage levels are equal to $2n+1$, where n denotes the cell count. However, THD can be minimized by the proper selection of switching angles. An advantage of this type of MLIs is that they require a smaller number of components than diode-clamped or flying capacitors. Therefore, the price and weight of these inverters are less than those of the other two inverters. An n-level cascaded H-bridge MLI is shown in Figure 1.25. It should be noted that the method of calculating the switching angles for this inverter is similar to the method used for previous MLIs.

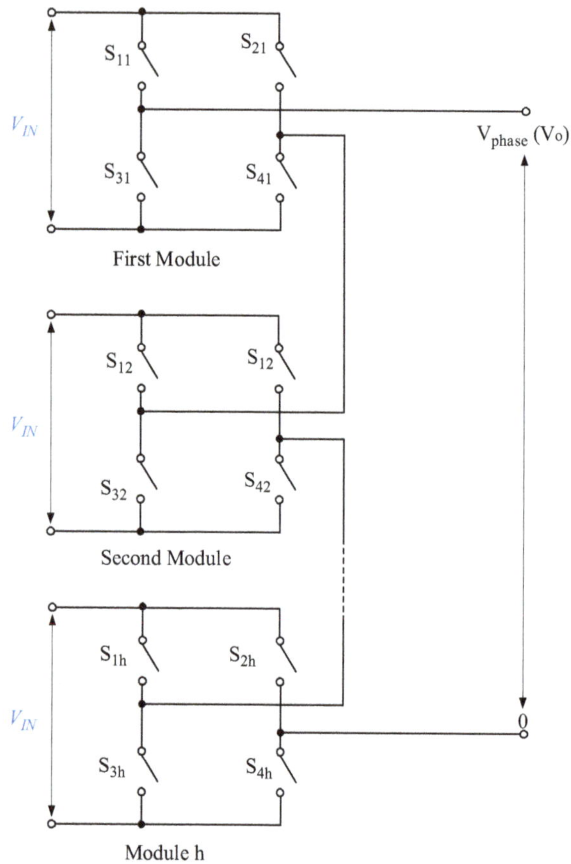

Figure 1.25 One phase of a cascaded H-bridge MLI

An n-level cascaded H-bridge MLI requires $2(n-1)$ switching devices, where n refers to the number of output voltage levels.

Five-level cascaded H-bridge MLIs

As shown for previous MLIs, the output voltage of a five-level cascaded H-bridge MLI contains five levels. Also, two cascaded H-bridge inverters are included in this inverter. Thus, eight switching devices are required by a five-level cascaded H-bridge MLI [33].

1.3.4 Review of a novel proposed MLIs

Various new MLI topologies with a small number of devices are discussed. Since three-phase structures act similarly to three single-phase structures with 120° shift of switching logic for each phase, the single-phase form is illustrated. For this case, valid switching states are also presented. Figure 1.26 shows a typical waveform of multilevel output voltage produced by MLIs to be used as a reference.

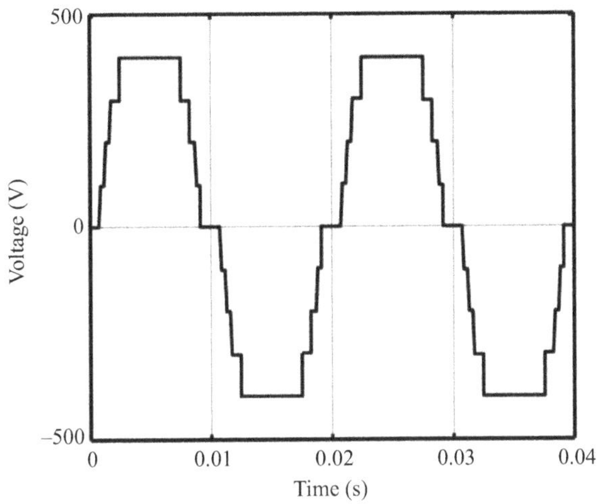

Figure 1.26 Multilevel output voltage waveform

1.3.4.1 Cascaded MLI with a reduced variety of DC voltage source

Reference [34] proposed a new MLI topology called reduced variety of DC voltage sources-based cascaded MLI (RVDC-CMLI). The proposed CMLI includes some basic cells connected in series. An H-bridge, bidirectional-conducting power switches, unidirectional-blocking power switches, and two symmetrical dc voltage sources are used in basic cells or modules. A "polarity generator" and a "level generator" are separately implemented for each cell. The proposed design has a fully modular structure, so more modules can be cascaded to extend the number of output DC levels. Figure 1.27 shows a nine-level MLI structure, which includes four input DC sources.

According to Figure 1.27, this topology includes no passive components. The only components that are employed in this topology are power switches, diodes, and voltage sources. The power switches of the polarity generator have to exhibit the lowest voltage blocking. This voltage is equal to the sum of the two voltage sources utilized in the corresponding module, i.e. ($V_{1.1}$ + $V_{1.2}$) for switches $T_{1.1}$, $T_{1.2}$, $T_{1.3}$, and $T_{1.4}$, while ($V_{2.1}$+ $V_{2.2}$) for switches $T_{2\cdot1}$, $T_{2\cdot2}$, $T_{2\cdot3}$, and $T_{2.4}$. These switches are high-rated compared to the switches used in the level generator. However, the power switches of the polarity generator can operate at fundamental switching frequency to allow fundamental frequency switching. Table 1.4 presents the valid combinations of switching operations to achieve different DC levels.

When a source is configured symmetrically, i.e., $V_{1.1} = V_{1.2} = V_{2.1} = V_{2.2} = V_{dc}$, multiple redundant states can be used to synthesize the voltage levels i.e. $\pm$ (V_{dc}, $2V_{dc}$, $3V_{dc}$, $4V_{dc}$). Accordingly, power can be shared among the cells. The proposed CMLI can flawlessly operate while the sources are configured symmetrically or asymmetrically. Different asymmetrical configurations have also been presented in

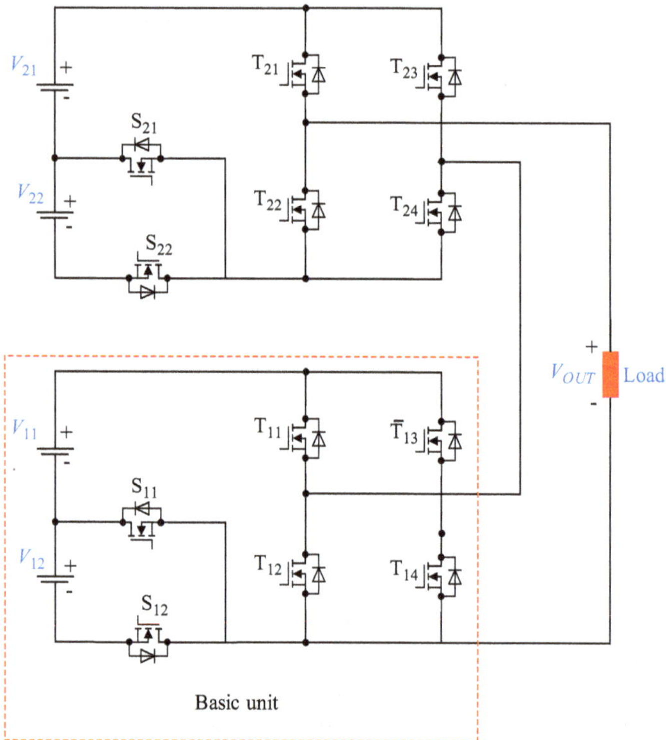

Figure 1.27 Cascaded MLI with reduced a variety of DC voltage sources

the literature. Therefore, the ratios of 1:2:4:8..., 1:4:16..., or 1:5:25 are feasible for sources in different modules. However, the magnitudes of the sources in the same module have to be identical. In a symmetrical configuration, a smaller number of switches are required for this topology compared to conventional symmetrical MLIs. Moreover, in asymmetrical configuration, the variety of the DC source magnitudes is smaller compared to the conventional asymmetrical MLIs.

1.3.4.2 Developed H-bridge topology

Reference [35] proposed the "developed H- bridge" for MLIs. This design benefits from the inherent features of H-bridge as well as asymmetrical sources. The sources are placed on the sides of the H-bridge to synthesize a multilevel output. Figure 1.28 shows this topology, which includes four DC sources. This design consists of an H-bridge around which bidirectional-conducting power switches, unidirectional-blocking power switches, and DC voltage sources are implemented. No distinct polarity generator or level exists. This means that the entire structure acts as a single unit without requiring passive components. This design is not modular, so the number of levels cannot be increased by repeating the same structure. However, in this topology, switches with various voltage ratings and

Table 1.4 Valid switching combination for RVDC-CMLI

State	V_{out}	ON-state switches
1	0	T_{11}, T_{13}, T_{21}, T_{23} T_{11}, T_{13}, T_{22}, T_{24} T_{12}, T_{14}, T_{22}, T_{24} T_{12}, T_{14}, T_{21}, T_{23}
2	V_{11}	S_{11}, T_{11}, T_{14}, T_{21}, T_{23} S_{11}, T_{11}, T_{14}, T_{22}, T_{24}
3	V_{21}	T_{11}, T_{13}, S_{21}, T_{21}, T_{24} T_{12}, T_{14}, S_{21}, T_{21}, T_{24}
4	$(V_{11}+V_{12})$	S_{12}, T_{11}, T_{14}, T_{21}, T_{23} S_{12}, T_{11}, T_{14}, T_{22}, T_{24}
5	$(V_{21}+V_{22})$	T_{11}, T_{13}, S_{22}, T_{21}, T_{24} T_{12}, T_{14}, S_{22}, T_{21}, T_{24}
6	$(V_{11}+V_{21})$	S_{11}, T_{11}, T_{14}, S_{21}, T_{21}, T_{24}
7	$(V_{11}-V_{12})$	S_{11}, T_{11}, T_{14}, S_{21}, T_{21}, T_{24}
8	$(V_{21}-V_{11})$	S_{11}, T_{21}, T_{13}, S_{21}, T_{22}, T_{23}
9	$(V_{11}+V_{12}) + V_{21}$	S_{12}, T_{11}, T_{14}, S_{21}, T_{21}, T_{24}
10	$(V_{11}+V_{12}) - V_{21}$	S_{12}, T_{11}, T_{14}, S_{21}, T_{22}, T_{23}
11	$V_{21}- (V_{11}+V_{12})$	S_{12}, T_{12}, T_{13}, S_{21}, T_{21}, T_{24}
12	$V_{11}+(V_{21}+V_{22})$	S_{11}, T_{11}, T_{14}, S_{22}, T_{21}, T_{24}
13	$V_{11}- (V_{21}+V_{22})$	S_{11}, T_{11}, T_{14}, S_{22}, T_{22}, T_{23}
14	$(V_{21}+V_{22}) - V_{11}$	S_{11}, T_{12}, T_{13}, S_{22}, T_{21}, T_{24}
15	$(V_{11}+V_{12})+ (V_{21}+V_{22})$	S_{12}, T_{11}, T_{14}, S_{22}, T_{21}, T_{24}
16	$(V_{11}+V_{12}) - (V_{21}+V_{22})$	S_{12}, T_{11}, T_{14}, S_{22}, T_{22}, T_{23}
17	$(V_{21}+V_{22}) - (V_{11}+V_{12})$	S_{12}, T_{12}, T_{13}, S_{22}, T_{21}, T_{24}
18	$-V_{11}$	S_{11}, T_{12}, T_{13}, T_{21}, T_{23} S_{11}, T_{12}, T_{13}, T_{22}, T_{24}
19	$-V_{21}$	T_{11}, T_{13}, S_{21}, T_{22}, T_{23} T_{12}, T_{14}, S_{21}, T_{21}, T_{24}
20	$-(V_{11}+V_{21})$	S_{12}, T_{11}, T_{14}, T_{21}, T_{23} S_{12}, T_{11}, T_{14}, T_{22}, T_{24}
21	$-(V_{21}+V_{22})$	T_{11}, T_{13}, S_{22}, T_{22}, T_{23} T_{12}, T_{14}, S_{22}, T_{22}, T_{23}
22	$-(V_{11}+V_{21})$	S_{11}, T_{12}, T_{13}, S_{21}, T_{22}, T_{23}
23	$-(V_{11}+V_{12}+V_{21})$	S_{12}, T_{12}, T_{13}, S_{21}, T_{22}, T_{23}
24	$-(V_{11}+V_{21}+V_{22})$	S_{11}, T_{12}, T_{13}, S_{22}, T_{22}, T_{23}
25	$(V_{11}+V_{12}) - (V_{21}+V_{22})$	S_{12}, T_{12}, T_{13}, S_{22}, T_{22}, T_{23}

sources with various magnitudes are required. The number of power switches is significantly reduced in this topology in comparison with the conventional cascaded H-bridges (CHB) at the same output voltage level. However, simultaneously, the voltage blocking is increased for different switch manifolds since the voltage sources are highly asymmetrical [35]. As Figure 1.28 shows, the voltage values of the switches have to be different, i.e., VR1 for SR1 & SR2, (VR2–VR1) for SR3 & SR4, VL1 for SL1 & SL2, (VL2 – V L1) for SL3 & SL4, and (VR2+ VL2) for SRa & SRb. Table 1.5 lists the possible switching states of the developed H-bridge for 25 levels.

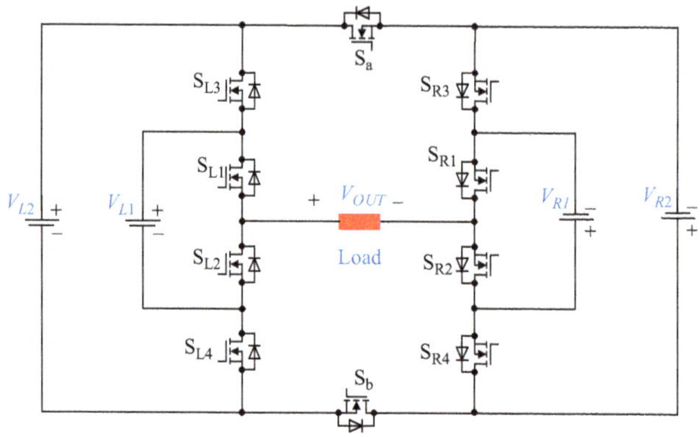

Figure 1.28 Developed H-bridge MLI

Table 1.5 Valid switching states for developed H-bridge

State	V_{out}	ON-state switches
1	$V_{L2}+V_{R2}$	S_{L1}, S_{L3}, S_{R1}, S_{R3}, S_b
2	$V_{L2}+V_{R2}-V_{L1}$	S_{L1}, S_{L3}, S_{R2}, S_{R3}, S_b
3	$V_{R2}+V_{L2}-V_{R1}$	S_{L2}, S_{L3}, S_{R2}, S_{R3}, S_b
4	$V_{L2}+V_{R2}-V_{L1}-V_{R1}$	S_{L2}, S_{L3}, S_{R2}, S_{R3}, S_b
5	$V_{L1}+V_{R2}$	S_{L1}, S_{L3}, S_{R1}, S_{R4}, S_b
6	V_{R2}	S_{L1}, S_{L3}, S_{R2}, S_{R4}, S_b
7	$V_{L1}+V_{R1}-V_{R2}$	S_{L2}, S_{L3}, S_{R1}, S_{R4}, S_b
8	$V_{R2}-V_{R1}$	S_{L2}, S_{L3}, S_{R2}, S_{R4}, S_b
9	$V_{L2}+V_{R1}$	S_{L1}, S_{L4}, S_{R1}, S_{R3}, S_b
10	$V_{L2}+V_{R1}-V_{L1}$	S_{L1}, S_{L4}, S_{R2}, S_{R3}, S_b
11	V_{L2}	S_{L2}, S_{L4}, S_{R1}, S_{R3}, S_b
12	$V_{L2}-V_{L1}$	S_{L2}, S_{L4}, S_{R2}, S_{R3}, S_b
13	$V_{L1}+V_{R1}$	S_{L1}, S_{L4}, S_{R1}, S_{R4}, S_b
14	V_{R1}	S_{L1}, S_{L4}, S_{R2}, S_{R4}, S_b
15	V_{L1}	S_{L2}, S_{L4}, S_{R1}, S_{R4}, S_b
16	0	S_{L2}, S_{L4}, S_{R1}, S_{R4}, S_b
		S_{L1}, S_{L3}, S_{R1}, S_{R3}, S_a
17	$-V_{L1}$	S_{L1}, S_{L3}, S_{R2}, S_{R3}, S_a
18	$-V_{R1}$	S_{L2}, S_{L3}, S_{R1}, S_{R3}, S_a
19	$-(V_{L1}+V_{R1})$	S_{L2}, S_{L3}, S_{R2}, S_{R3}, S_a
20	$-(\mathrm{V}_{L2}-_{L1})$	S_{L1}, S_{L3}, S_{R1}, S_{R4}, S_a
21	$-V_{L2}$	S_{L1}, S_{L3}, S_{R2}, S_{R4}, S_a
22	$-(V_{L2}+V_{R1}-V_{L1})$	S_{L2}, S_{L3}, S_{R1}, S_{R4}, S_a
23	$-(\mathrm{V}_{L2}+-_{R1})$	S_{L1}, S_{L4}, S_{R1}, S_{R3}, S_a
24	$-(V_{R2}-V_{R1})$	S_{L2}, S_{L3}, S_{R2}, S_{R4}, S_a
25	$-(\mathrm{V}_{L1}-V_{R1}-V_{R2})$	S_{L1}, S_{L4}, S_{R2}, S_{R3}, S_a
26	$-V_{R2}$	S_{L2}, S_{L3}, S_{R2}, S_{R4}, S_a
27	$-(V_{L1}+-_{R2})$	S_{L2}, S_{L4}, S_{R2}, S_{R3}, S_a
28	$-(V_{L2}+V_{R2}-V_{L1}-V_{R1})$	S_{L1}, S_{L4}, S_{R1}, S_{R4}, S_a
29	$-(V_{L2}+V_{R2}-V_{R1})$	S_{L1}, S_{L4}, S_{R2}, S_{R4}, S_a
30	$-(V_{L2}+V_{R2}-V_{L1})$	S_{L2}, S_{L4}, S_{R1}, S_{R4}, S_a
31	$-(\mathrm{V}_{L2}+V_{R2})$	S_{L2}, S_{L4}, S_{R2}, S_{R4}, S_a

1.3.4.3 Switched capacitor unit (SCU)-based MLI

The SCU-based MLI, which is a multilevel topology, was presented in Ref. [36]. This inverter has 17 levels and four sources as illustrated in Figure 1.29. In this inverter, the basic cells are cascaded to obtain different levels, and an H-bridge is used to reverse polarity. Four bidirectional-conducting/unidirectional-blocking power switches, a capacitor, a diode, and a DC source are used in each modular cell. An additional level is generated using the capacitor. The switches initially charge the capacitor, and then the capacitor can be considered as an additional level. Therefore, this topology can be assumed as an MLI with a switched-capacitor unit.

The level generator in this topology is distinct with sources (V_1, V_2, V_3, & V_4), switches (S_1, P_1, B_1, Z_1, S_2, P_2, B_2, Z_2, S_3 P_3, B_3, Z_3, S_4, P_4, B_4, & Z_4), capacitors (C_1, C_2, C_3, & C_4) and diodes (D_1, D_2, D_3, & D_4). Moreover, the H-bridge of the polarity generator includes four power switches (T_1, T_2, T_3, & T_4). In this topology,

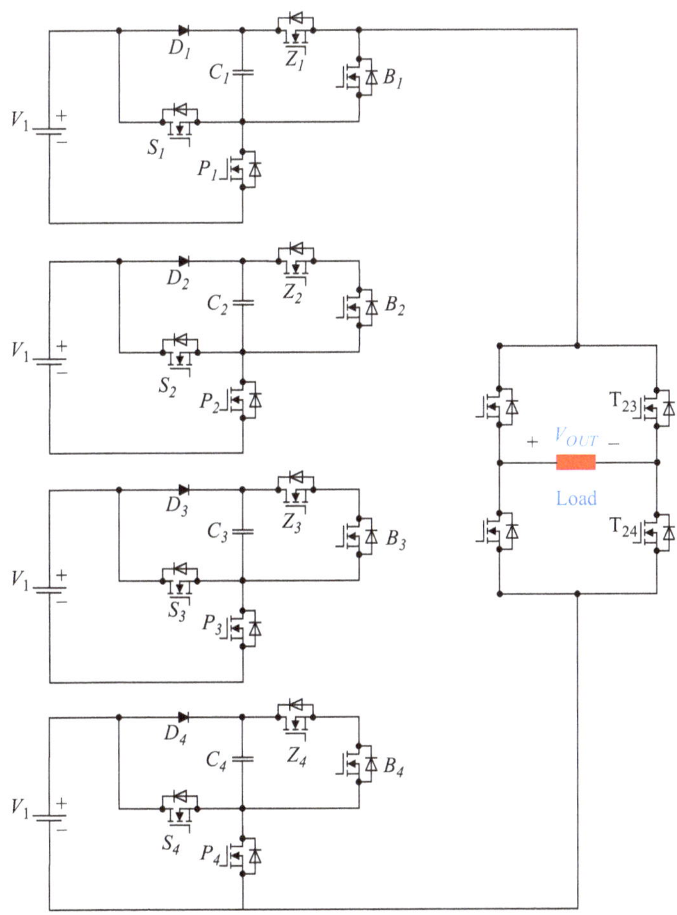

Figure 1.29 An SC-based MLI

the number of switches and driver circuits is significantly reduced compared to the conventional CHB for the same output voltage levels. However, the presence of the capacitor leads to some issues in dynamic performance.

As shown in Figure 1.29, the switches of the level generator i.e., S_1, P_1, B_1, Z_1, S_2, P_2, B_2, Z_2, S_3 P_3, B_3, Z_3, S_4, P_4, B_4, & Z_4 should possess the capability of minimum voltage blocking, which is equal to the source voltage magnitude in the basic cell of the corresponding switch. The polarity generator includes the switches T_1, T_2, T_3, and T_4 with minimum voltage blocking, which is twice the sum of all source voltages i.e., $2(V_1+ V_2+ V_3+ V_4)$. Table 1.6 presents the valid combinations of switching operations to achieve different DC levels.

The operation frequency in this design can be the fundamental switching frequency. If the source is configured symmetrically, the voltage levels (V_{dc}, $2V_{dc}$, $3V_{dc}$, $4V_{dc}$) can be realized by combining the input sources. Therefore, an appropriate control scheme can be used to equally share power among the cells. This design can be used for both symmetrically and asymmetrically configured sources. Two asymmetrical source configurations are proposed in this study, which require an extra diode and switches for each cell. Moreover, another topology with slight modification in the basic module is proposed in this study. Contrary to conventional designs that use a single H-bridge to invert the output voltage polarity and cascade only capacitor-based modules, in this study, H-bridge is integrated with each of the capacitor-based modules, and the obtained structures are then cascaded. The number of switches in the second topology is larger than that in the first one, while the stress of the devices in the H-bridge is reduced.

1.3.4.4 Proposed six-level NPC-based grid-tied inverter

The electrical equivalent circuit of the suggested structure of [37] has been indicated in Figure 1.30. Regarding Figure 1.30, the introduced structure utilizes six

Table 1.6 Valid switching states for developed H-bridge

State	V_{out}	ON-state switches
1	0	B_1, P_1, B_2, P_2, B_3, P_3, B_4, P_4
2	V_1	Z_1, P_1, B_2, P_2, B_3, P_3, B_4, P_4
3	$2V_1$	S_1, Z_1, B_2, P_2, B_3, P_3, B_4, P_4
4	V_2	B_1, P_1, Z_2, P_2, B_3, P_3, B_4, P_4
5	$2V_2$	B_1, P_1, S_2, Z_2, B_3, P_3, B_4, P_4
6	V_3	B_1, P_1, B_2, P_2, Z_3, P_3, B_4, P_4
7	$2V_3$	B_1, P_1, B_2, P_2, S_3, Z_3, B_4, P_4
8	V_4	B_1, P_1, B_2, P_2, B_3, P_3, Z_4, P_4
9	$2V_4$	B_1, P_1, B_2, P_2, B_3, P_3, S_4, Z_4
10	V_1+V_2	Z_1, P_1, Z_2, P_2, B_3, P_3, B_4, P_4
11	$2V_1+V_2$	S_1, Z_1, Z_2, P_2, B_3, P_3, B_4, P_4
12	V_1+2V_2	Z_1, P_1, Z_2, P_2, B_3, P_3, B_4, P_4
13	$2V_1+2V_2$	S_1, Z_1, Z_2, P_2, B_3, P_3, B_4, P_4
14	V_1+V_3	Z_1, P_1, B_2, P_2, Z_3, P_3, B_4, P_4

(Continues)

Table 1.6 (Continued)

State	V_{out}	ON-state switches
15	$2V_1+V_3$	S_1, Z_1, B_2, P_2, Z_3, P_3, B_4, P_4
16	V_1+2V_3	Z_1, P_1, B_2, P_2, S_3, Z_3, B_4, P_4
17	$2V_1+2V_3$	S_1, Z_1, B_2, P_2, S_3, Z_3, B_4, P_4
18	V_1+V_4	Z_1, P_1, B_2, P_2, B_3, P_3, Z_4, P_4
19	$2V_1+V_4$	S_1, Z_1, B_2, P_2, B_3, P_3, Z_4, P_4
20	V_1+2V_4	Z_1, P_1, B_2, P_2, B_3, P_3, S_4, P_4
21	$2V_1+2V_4$	S_1, Z_1, Z_2, P_2, B_3, P_3, S_4, Z_4
22	V_2+V_3	B_1, P_1, Z_2, P_2, Z_3, P_3, B_4, P_4
23	V_2+2V_3	B_1, P_1, Z_2, P_2, S_3, Z_3, B_4, P_4
24	V_2+2V_3	B_1, P_1, Z_2, P_2, S_3, Z_3, B_4, P_4
25	V_2+V_4	B_1, P_1, Z_2, P_2, B_3, P_3, Z_4, P_4
26	$2V_2+V_4$	B_1, P_1, Z_2, P_2, B_3, P_3, Z_4, P_4
27	V_2+2V_4	B_1, P_1, Z_2, P_2, B_3, P_3, S_4, P_4
28	V_2+2V_4	B_1, P_1, Z_2, P_2, B_3, P_3, S_4, Z_4
29	V_3+V_4	B_1, P_1, B_2, P_2, Z_3, P_3, Z_4, P_4
30	$2V_3+V_4$	B_1, P_1, B_2, P_2, S_3, Z_3, Z_4, P_4
31	V_3+2V_4	B_1, P_1, B_2, P_2, Z_3, P_3, S_4, Z_4
32	$2V_3+2V_4$	B_1, P_1, B_2, P_2, S_3, Z_3, S_4, Z_4
33	$V_1+V_2+V_3$	Z_1, P_1, Z_2, P_2, Z_3, P_3, B_4, P_4
34	$2V_1+V_2+V_3$	S_1, Z_1, Z_2, P_2, Z_3, P_3, B_4, P_4
35	$V_1+2V_2+V_3$	Z_1, P_1, Z_2, P_2, Z_3, P_3, B_4, P_4
36	$V_1+V_2+2V_3$	Z_1, P_1, Z_2, P_2, S_3, Z_3, B_4, P_4
37	$2V_1+2V_2+V_3$	S_1, Z_1, Z_2, P_2, Z_3, P_3, B_4, P_4
38	$2V_1+V_2+2V_3$	S_1, Z_1, Z_2, P_2, S_3, Z_3, B_4, P_4
39	$V_1+2V_2+2V_3$	Z_1, P_1, Z_2, P_2, S_3, Z_3, B_4, P_4
40	$2V_1+2V_2+2V_3$	S_1, Z_1, Z_2, P_2, S_3, Z_3, B_4, P_4
41	$V_2+V_3+V_4$	B_1, P_1, Z_2, P_2, Z_3, P_3, Z_4, P_4
42	$2V_2+V_3+V_4$	B_1, P_1, Z_2, P_2, Z_3, P_3, Z_4, P_4
43	$V_2+2V_3+V_4$	B_1, P_1, Z_2, P_2, B_3, P_3, Z_4, P_4
44	$V_2+V_3+2V_4$	B_1, P_1, Z_2, P_2, Z_3, P_3, S_4, Z_4
45	$2V_2+2V_3+V_4$	B_1, P_1, Z_2, P_2, S_3, Z_3, Z_4, P_4
46	$V_2+2V_3+2V_4$	B_1, P_1, Z_2, P_2, S_3, Z_3, S_4, Z_4
47	$2V_2+V_3+2V_4$	B_1, P_1, Z_2, P_2, S_3, P_3, S_4, Z_4
48	$2V_2+2V_3+2V_4$	B_1, P_1, Z_2, P_2, S_3, Z_3, S_4, Z_4
49	$V_1+V_3+V_4$	Z_1, P_1, B_2, P_2, Z_3, P_3, Z_4, P_4
50	$2V_1+V_3+V_4$	S_1, Z_1, B_2, P_2, Z_3, P_3, Z_4, P_4
51	$V_1+2V_3+V_4$	Z_1, P_1, B_2, P_2, S_3, Z_3, Z_4, P_4
52	$V_1+V_3+2V_4$	Z_1, P_1, B_2, P_2, Z_3, P_3, S_4, Z_4
53	$2V_1+2V_3+V_4$	S_1, Z_1, B_2, P_2, S_3, Z_3, Z_4, P_4
54	$V_1+2V_3+2V_4$	Z_1, P_1, B_2, P_2, S_3, Z_3, S_4, Z_4
55	$2V_1+V_3+2V_4$	S_1, Z_1, B_2, P_2, Z_3, P_3, S_4, Z_4
56	$2V_1+V_3+2V_4$	S_1, Z_1, B_2, P_2, S_3, Z_3, S_4, Z_4
57	$V_1+V_2+V_4$	Z_1, P_1, Z_2, P_2, B_3, P_3, Z_4, P_4
58	$2V_1+V_2+V_4$	S_1, Z_1, Z_2, P_2, B_3, P_3, Z_4, P_4
59	$V_1+2V_2+V_4$	Z_1, P_1, Z_2, P_2, B_3, P_3, Z_4, P_4
60	$V_1+V_2+2V_4$	Z_1, P_1, Z_2, P_2, B_3, P_3, S_4, Z_4
61	$2V_1+2V_2+V_4$	S_1, Z_1, Z_2, P_2, B_3, P_3, Z_4, P_4
62	$V_1+2V_2+2V_4$	Z_1, P_1, Z_2, P_2, B_3, P_3, S_4, Z_4
63	$2V_1+V_2+2V_4$	S_1, Z_1, Z_2, P_2, B_3, P_3, S_4, Z_4

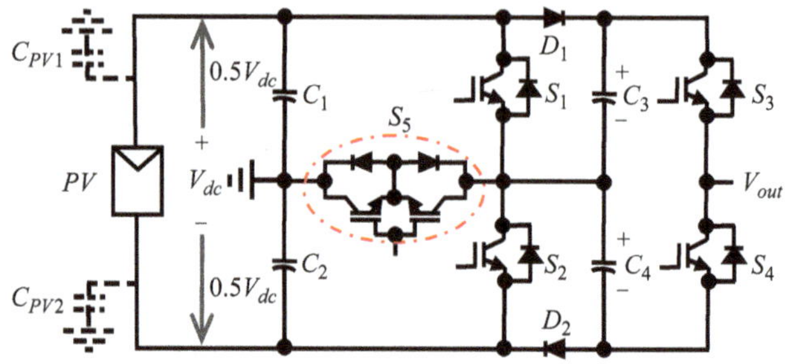

Figure 1.30 Six-level NPC-based grid-tied inverter [37]

switches, two diodes, and four capacitors. Two switch-leg are used in the mentioned topology. Switches S_1, S_2 and S_3, S_4 are involved in the first and second switch legs, respectively. The capacitors C_1 and C_2 are charged to the half value of the power supply (V_{C1}=V_{C2}=0.5 V_{dc}). Note that, in this inverter, the capacitors C_3 and C_4 are charged to the input dc source (V_{C3}=V_{C4}=V_{dc}). With respect to Figure 1.30, in the proposed inverter, the midpoint of the DC-link is connected to the null of the grid, directly. As a result, the CMV can be fixed to the constant value and the leakage current can be reduced.

With respect to Figure 1.30, the capacitor's voltage V_{C1} and V_{C2} can be provided as:

$$V_{C1} = 0.5V_{dc} - \Delta V_{C1}(t) \tag{1.1}$$

$$V_{C2} = 0.5V_{dc} - \Delta V_{C2}(t) \tag{1.2}$$

$$\Delta V_{C1}(t) = -\Delta V_{C2}(t) = \frac{I_{mg}\cos(\omega t)}{\omega(C_1 + C_2)} \tag{1.3}$$

In the above-mentioned equations, I_{mg} denotes the maximum value grid current waveform. Regarding (1.3), the capacitor's voltage includes DC and AC components. Considering ($|\Delta V_{C1}| = |\Delta V_{C2}| = |\Delta V_C|$), the leakage current can be calculated as follows:

$$I_{Leakage} = (C_{PV1} + C_{PV2})\frac{d(\Delta V_C)}{dt} \tag{1.4}$$

By replacing (1.2) into (1.4), the final value of the leakage current equation is obtained :

$$I_{Leakage} = \frac{(C_{PV1} + C_{PV2})}{(C_1 + C_2)} I_{mg}\sin(\omega t) \tag{1.5}$$

Considering (1.5), it is clear that the leakage current is varied only by the AC term. It should be noted that the capacitors C_{pv1} and C_{pv2} are smaller than C_1 and C_2, and because the AC term oscillates with grid frequency, the leakage current is close to zero. The suggested structure has some benefits such as limitation of leakage current, supporting both active and reactive powers, and voltage boosting. The suggested inverter has a high step-up voltage gain inherently. In the mentioned solution, in order to adjust both active and reactive powers, the peak current control (PCC) method has been implemented. By using the PCC technique, the grid current quality is improved and it has a suitable quality at different power factors.

1.3.4.5 Proposed seven-level NPC-based grid-connected inverter

The introduced NPC-based seven-level grid-connected PV inverter which is provided in [38] is indicated in Figure 1.31. With respect to Figure 1.31, the suggested inverter uses six switches, four capacitors, and two diodes to generate a seven-level output voltage. In the mentioned topology, the DC-link capacitors ($C_{dc\text{-}1}$ and $C_{dc\text{-}2}$) are fixed to the $0.5V_{dc}$ ($V_{dc\text{-}1}=V_{dc\text{-}2}=0.5V_{dc}$). It should be noted that the switched capacitors C_1 and C_2 are adjusted to input dc power supply ($V_{C1}=V_{C2}=V_{dc}$).

The suggested structure provides a voltage boosting feature with a boosting factor of 1.5. In addition, the proposed topology can limit the leakage current using the NPC strategy.

1.3.4.6 Five-level NPC-based grid-connected inverter

The suggested 5-level inverter of [39] is presented in Figure 1.32. In this structure, in order to produce a five-level output voltage, six unidirectional switches, and two

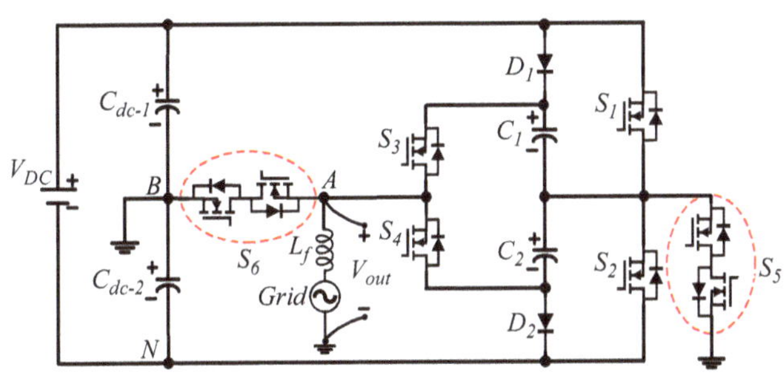

Figure 1.31 Proposed seven-level topology [38]

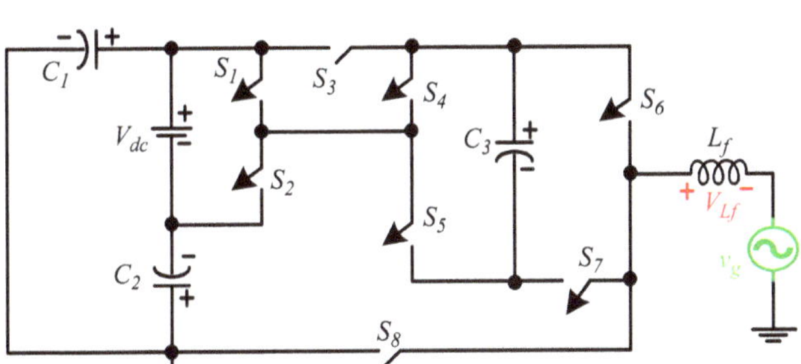

Figure 1.32 Proposed five-level topology [39]

bidirectional switches, three capacitors (C_1–C_3) are applied. Note that the DC-link capacitors are fixed to half of the input dc supply (V_{C1}=V_{C2}=0.5V_{dc}) by charging/discharging switched capacitor strategy. Also, the switched capacitor is adjusted to input dc source (V_{C3}=V_{dc}). In this structure, by charging and discharging the used capacitors, the output voltage levels are generated. With respect to Figure 1.32, in the suggested inverter, the midpoint of DC-link capacitors is tied to the null of the grid, directly. So, the leakage current is limited by this solution in the PV-based power generation systems. Also, by this solution, the grid current THD is improved and a sinusoidal current waveform is injected into the grid. The proposed topology uses a switched-capacitor (SC) cells to generate five-level output voltage waveforms with a voltage boosting feature and voltage gain of 1.5.

In addition, the voltage of used capacitors is adjusted inherently and there is a need to control the system for balancing the capacitor's voltage.

Table 1.7 summarizes the comparison results of a comparison study between presented multilevel structures. In addition to the mentioned topologies, new multilevel topologies were presented recently. Although some of the suggested solutions provide the benefit of the reduced number of passive elements and semiconductors, they do not have good and accurate performance in terms of control, voltage adjusting strategies, fault tolerance, and reliability. As a result, these structures have not been applied in the literature for this application. Some of the new solutions with their benefits are summarized in Table 1.8.

Further, cost reduction, as an important goal for electrified transport in the future, plays an important role in the selection of the inverter structure. The use of multi-level solutions affects operating costs. Regarding the side effects of these structures, variations are not limited to the converter. In the following, cost changes in the topology and other units are considered. A comparative cost study is considered between a two-level IGBT-based pull-up structure and a CHB solution in [41]. Apart from the cost of inverter structure, the cost of storing in the battery is considered by applying the multi-level topology. Since the multilevel topology has been shown to be more efficient in the same solution, the vehicle can maintain the same driving distance with a lower battery capacity. The comparison results are summarized in Table 1.9 [41]. It can be understood that although the inverter cost is higher in the case of CHB solution, the total system cost has been slightly mitigated by implementing a multilevel structure. Reference [42] provides another comparison study between a conventional two-level topology and three-level NPC-based solution. Note that in both structures, the switching devices are IGBTs. Table 1.10 presents the costs of IGBTs, voltage sensors, fast diodes, gate drivers, control systems, used capacitors, heat sinks, and filters. In this comparison, the cost of the battery is not considered. So, it can be suitable for grid-tied electrified transportation. It can be understood that by reducing the cost of the filter, the overall cost of the system is decreased by 3% of the three-level NPC-based structure.

Table 1.7 A comparison of multilevel inverter structures

Topology	Advantages	Disadvantages
Neutral point clamped (NPC)	• No floating capacitor • Satisfied dynamic response • Simple structure • Efficient cost	• Increasing cost and decreasing reliability by increasing the number of generating voltage levels • Non-uniformity of switches losses
Active neutral point clamped (ANPC)	• Satisfied dynamic response • Simple structure • The same loss in the switches • Efficient cost	• Increasing the number of floating capacitors with increasing voltage level • Have intricate control system compared with NPC topology • Various voltage rating of switches at higher levels
Flying capacitor (FC)	Efficient cost	• The large count of floating capacitors • Intricate control system • Unacceptable voltage ripple at low frequencies • High volume and low power density • Using voltage sensors • Low reliability
Cascaded H-bridge (CHB)	• Have a good reliability • Modular • No floating capacitors • Simple control system	• Using isolated Input DC power supply which is not suitable for PV applications • Use more count of power switches
Modular multilevel converter (MMC)	• Have a good reliability • Modular • Scalable	• Use large count of capacitors • Intricate control system • Pre-charging circuits for the capacitors • Unacceptable voltage ripple at low frequencies • High volume and low power density • Use a lot of voltage sensors
Nested neural point clamped (NNPC)	• Use less count of capacitors compared with FC • Use less count of diodes compared with NPC	• All of the issues related to floating capacitors • Unequal voltage stress on switches at levels higher than 4
T-Type multi-level inverter	• Not using capacitors • Not using diodes • Simple control strategy	• Higher voltage stress on switches • Low efficiency at high frequency operation due to high switching losses • Not suitable for high-voltage applications

Table 1.8 New multilevel topologies: benefits and disadvantages [40]

Topology	Advantages	Drawbacks
Switched-capacitor topology	• Reduced switches at high number of output voltage levels • Modular structure and equal load sharing among power supplies	• Used isolated input DC supply • Used large count of capacitors
Extended cascaded topology	• Switching frequency is equal to fundamental frequency • High modularity and simplicity	• Used isolated DC power supply • Unequal power sharing among cells • Different voltage stress on switches
Cascaded power circuit topology	• Having the same loss in the switches • Very simple topology • Use non-isolated DC power supply	• Different voltage stress on switches
Novel cascaded-based structure	• Have low power switching losses and high overall efficiency	• Different power sharing • Used isolated DC supplies
Hybrid structure	• Equal power sharing • Good choice for all voltage levels • Switching frequency equal to fundamental frequency	• Different voltage stress on switches • Complex control system • Used isolated DC supplies
Highly efficient and reliable cascaded topology	• Very simple structure • Equal power sharing • Low power losses and high efficiency	• Used isolated DC supplies
Cross-connected source-based inverter	• Very simple structure • Switching frequency is equal to fundamental frequency • Suitable for all output voltage levels	• Used asymmetric voltage power supplies • Used bidirectional switches • Unequal power sharing
Multilevel cascade inverter	• Very simple topology • Low conduction losses • Suitable for all voltage levels	• Different voltage stress on switches • Unequal power sharing • Used bidirectional switches
Reduced component CHB-based inverter	• Very simple structure at high levels • High overall efficiency	• Used isolated voltage supplies • Used bidirectional switches

(Continues)

Table 1.8 (*Continued*)

Topology	Advantages	Drawbacks
		• Unequal power sharing • Different voltage stress on switches
Cascaded unit-based inverter	• Switching frequency is equal to low fundamental frequency	• Different voltage stress on switches • Unequal power sharing • Used isolated voltage supplies

Table 1.9 *Cost comparison based on [41]*

Costs	Two-level topology	Seven-level CHB structure
Switches	296.6 ($)	203.4 ($)
Gate drivers	39 ($)	234 ($)
Capacitors	30.3 ($)	0 ($)
Heatsink	17.3 ($)	16.2 ($)
Controller	150 ($)	300 ($)
Battery	5,760 ($)	5,310 ($)
Contactor	74 ($)	180 ($)
Overhead cost	133.3 ($)	188.6 ($)
Total cost	6,500.5 ($)	6,433.3 ($)

Table 1.10 *Cost comparison based on [42]*

Costs	Two-level structure	Three-level NPC structure
Switches	204 ($)	216 ($)
Fast diodes	0 ($)	72 ($)
Capacitors	90 ($)	90 ($)
Heatsink	400 ($)	400 ($)
Controller	150 ($)	300 ($)
Gate drivers	60 ($)	120 ($)
Filter	842 ($)	640 ($)
Sensors	80 ($)	90 ($)
Total	1,726 ($)	1,678 ($)

1.4 Conclusion

In this chapter, different fundamental topologies for non-isolated and isolated converters have been reviewed. These topologies include buck–boost, SEPIC, cuck, Z-source, and zeta for non-isolated converters and fly-back, bridge, and push–pull converters for isolated converters. These converters have been comprised based on their advantages and disadvantages. In addition, state-of-the-art of non-isolated converter topologies and resonant converters have been represented. Also, different kinds of multilevel inverters have been reviewed and studied. In recent years, due to the lack of fossil energy sources, the use of renewable energy sources such as solar energy has increased significantly. PV inverters are the most important devices in the PV systems. Transformer-less grid-tied inverters have been expanded for many years and have had good applications in grid-connected PV systems. On the other hand, inverters without transformers are very important for achieving the proper voltage gain from relatively low solar panels and providing the power grid's proper voltage peak. In this chapter, some NPC-based and common grounded grid-tied inverters are reviewed. In addition, the switching states of the switches and charging states of the capacitors in the reviewed topologies have been considered. In order to show the benefits and drawbacks of each presented inverter, a comprehensive comparison is done.

References

[1] Gopi, A. and Saravanakumar, R. 'High step-up isolated efficient single switch DC–DC converter for renewable energy source'. *Ain Shams Engineering Journal*, 2014, 5(4): 1115–1127.

[2] Rajabi, A., Rajaei, A., Tehrani, V.M., *et al.* 'A non-isolated high step-up DC–DC converter using voltage lift technique: analysis, design, and implementation'. *IEEE Access* 2022, 10: 6338–6347.

[3] Yusivar, F., Farabi, M.Y., Suryadiningrat, R., Ananduta, W.W., and Syaifudin, Y. 'Buck-converter photovoltaic simulator'. *International Journal of Power Electronics and Drive Systems (IJPEDS)*. 2011, 1(2): 156–167.

[4] Rajabi, A., Mohammadzadeh Shahir, F., and Babaei, E. 'Designing a novel voltage-lift technique based non-isolated boost DC–DC converter with high voltage gain'. *International Transactions on Electrical Energy Systems.* 2021, 13: e13213.

[5] Mohan, N. *Power Electronics and Drives.* Indian Institute of Technology' Guwahati, India, 2003.

[6] Bose, B.K. 'Power electronics, smart grid, and renewable energy systems'. *Proceedings of the IEEE*, 2017, 105(11): 2011–2018.

[7] Zhu, M. and Luo, F.L. 'Enhanced self-lift Cûk converter for negative-to-positive voltage conversion'. *IEEE Transactions on Power Electronics.* 2010, 25(9): 2227–2233.

[8] Babaei, E., Abu-Rub, H., and Suryawanshi, H.M. 'Z-source converters: topologies, modulation techniques, and application – Part I'. *IEEE Transactions on Industrial Electronics*, 2018, 65(6): 5092–5095.

[9] Wu, T.F., Liang, S.A., and Chen, Y.M. 'Design optimization for asymmetrical ZVS-PWM Zeta converter'. *IEEE Transactions on Aerospace and Electronic Systems*, 2003, 39(2): 521–532.

[10] Vikram, AA., Navaneeth, R., Kumar, M.N., and Vinoth, R. 'Solar PV array fed BLDC motor using zeta converter for water pumping applications'. *IEEE Transactions on Industry Applications*, 2018, 3(4): 8–19.

[11] Keerthana, R. and Chintu, N.J. 'Performance analysis of zeta converter in wind power application'. *Asian Journal of Applied Science and Technology (AJAST)*, 2017, 1(3): 199–203.

[12] Rajabi, A.R., Mohammadadeh Shahir, F., and Sedaghati, R. 'New unidirectional step-up DC–DC converter for fuel–cell vehicle: design and implementation'. *Electric Power System Research*, 2022, 212: 108653.

[13] Hu, X. and Gong, C. 'A high gain input-parallel output-series DC/DC'. *IEEE Transactions on Power Electronics*, 2015, 30 (3): 1306–1317.

[14] Bhaskar, M.S., Meraj, M., Iqbal, A., Padmanaban, S., Maroti, P.K., and Alammari, R. 'High gain transformer-less double-duty-triple-mode DC/DC converter for DC microgrid'. *IEEE Access*, 2019, 7: 36353–36370.

[15] Maroti, P.K., Padmanaban, S., Bhaskar, M.S., Meraj, M., Iqbal, A., and Al-Ammari, R. 'High gain three-state switching hybrid boost converter for DC microgrid applications'. *IET Power Electronics*, 2019, 12(14), 3656–3667.

[16] Sukesh, N., Pahlevaninezhad, M., and Jain, P.K. 'Analysis and implementation of a single stage flyback PV microinverter with soft switching'. *IEEE Transactions on Industrial Electronics*, 2014, 61(4): 1819–1833.

[17] Wu, H., Chen, R., Zhang, J., Xing, Y., Hu, H., and Ge, H. 'A family of three-port half-bridge converters for a stand-alone renewable power system'. *IEEE Transactions on Power Electronics*. 2011, 26(9): 2697–2706.

[18] Hu, W., Wu, H., Xing, Y., and Sun, K. 'A full-bridge three-port converter for renewable energy application'. In *Proceedings of the 2014 IEEE Applied Power Electronics Conference and Exposition-APEC 2014*, Fort Worth, TX, 2014.

[19] Jeong, D.K., Kim, H.S., Baek, J.W., Kim, J.Y., and Kim, H.J. 'Dual active bridge converter for Energy Storage System in DC microgrid'. In *Proceedings of the 2016 IEEE Transportation Electrification Conference and Expo, Asia-Pacific (ITEC Asia-Pacific 2016)*, Busan, Korea, 2016.

[20] Petit, P., Aillerie, M., Sawicki, J.P., and Charles, J.P. 'Push-pull converter for high efficiency photovoltaic conversion'. *Energy Procedia*, 2012, 18: 1583–1592.

[21] Steigerwald, R. 'A comparison of half-bridge resonant converter topologies'. *IEEE Transactions on Power Electronics*, 1988, 3 (2): 174–182.

[22] Lazar, J.F. and Martinelli, R. 'Steady-state analysis of the LLC series resonant converter'. In *APEC 2001. Sixteenth Annual IEEE Applied Power*

Electronics Conference and Exposition (Cat. No. 01CH37181), IEEE, 2001, vol. 2, pp. 728–735.

[23] Kjaer, S.B., Pedersen, J.K., and Blaabjerg, F. 'A review of single-phase grid-connected inverters for photovoltaic modules'. *IEEE Transactions on Industry Applications*, 2005, 41(5): 1292–1306.

[24] Freddy, T.K.S., Rahim, N.A., Hew, W.P., and Che, H.S. 'Modulation techniques to reduce leakage current in three phase tranformerless H7 photovoltaic inverter'. *IEEE Transactions on Industrial Electronics*, 2015, 62: 322–331.

[25] Tran, T.T., Nguyen, M.K., Duong, T.D., Choi, J.H., Lim, Y.C., and Zare, F. 'A switched capacitor voltage doubler based boost inverter for common mode voltage reduction'. *IEEE Access*, 2019, 7: 98618–98629.

[26] Peng, F.Z. 'Z-source inverter'. *IEEE Transactions on Industry Applications*, 2003, 39: 504–510.

[27] Siwakoti, Y.P., Peng, F.Z., Blaabjerg, F., Loh, P.C., Town, G.E. 'Impedance-source networks for electric power conversion. Part I: a topological review'. *IEEE Transactions on Power Electronics*, 2015, 30: 699–716.

[28] Erginer, V. and Sarul, M.H. 'A novel reduced leakage-current modulation technique for Z source inverter used in photovoltaic systems'. *IET Power Electronics*, 2014, 7: 496–502.

[29] Duong, T.-D., Nguyen, M.-K., Tran, T.-T., Vo, D.-V., Lim, Y.-C., and Choi, J.-H. 'Topology review of three-phase two-level transformerless photovoltaic inverters for common-mode voltage reduction'. *Energies*, 2022, 15: 3106.

[30] Duong, T.D., Nguyen, M.K., Tran, T.T., Lim, Y.C., Choi, J.H., and Wang, C. 'Modulation techniques for a modified three-phase quasi-switched boost inverter with common-mode voltage reduction'. *IEEE Access*, 2020, 8: 160670–160683.

[31] Fang, Z. and Peng, A. 'Generalized MLI topology with self voltage balancing'. *IEEE Transaction on Industry applications*, 2001, 37(2): 611–618.

[32] Tolbert, L.M., Chiasson, J.N., and McKenzie, K.J. 'Elimination of harmonics in a multilevel converter with non-equal DC sources'. *IEEE Transaction on Industry applications*. 2005, 41(1): 75–82.

[33] Rodríguez, J., Lai, J.S., and Peng, F. 'MLIs: a survey of topologies, controls and applications'. *IEEE Transaction on Industry Electronics*. 2002, 49(4): 724–738.

[34] Bahravar, S., Babaei, E., and Hosseini, S.H. 'New cascaded multilevel inverter topology with reduced variety of magnitudes of DC voltage sources'. In *Proceeding of the IEEE 5th India International Conference on Power Electronics (IICPE)*, New Delhi, India, 2012, pp. 1–6.

[35] Babaei, E., Alilu, S., and Laali, S. 'A new general topology for cascaded multilevel inverters with reduced number of components based on developed H-bridge'. *IEEE Transactions on Industrial Electronics*, 2014, 61(8): 3932–3939.

[36] Babaei, E. and Gowgani, S.S., 'Hybrid multilevel inverter using switched capacitor units'. *IEEE Transactions on Industrial Electronics*, 2014, 61(9): 4614–4612.

[37] Kurdkandi, N.V., Marangalu, M.G., Mohammadsalehian, S., Tarzamni, H., and Siwakoti, Y.P., 'A new six-level transformer-less grid-connected solar photovoltaic inverter with less leakage current'. *IEEE Access*, 2022, 10: 63736–63753.

[38] Kurdkandi, N.V., Marangalu, M.G., Husev, O., and Aghaei, A., 'A new seven-level transformer-less grid-tied inverter with leakage current limitation and voltage boosting feature'. *IEEE Journal of Emerging and Selected Topics in Industrial Electronics*, 2022, 4: 1–11.

[39] Kurdkandi, N.V., Marangalu, M.G., Hemmati, T., Mehrizi-Sani, A., Rahimpour, S., and Babaei, E., 'Five-level NPC based grid-tied inverter with voltage boosting capability and eliminated leakage current'. In *Proceedings of the 2022 13th Power Electronics, Drive Systems, and Technologies Conference (PEDSTC)*, IEEE, 2022, pp. 676–680.

[40] Omer, P., Kumar, J., and Surjan, B.S., 'A review on reduced switch count multilevel inverter topologies'. *IEEE Access*, 2020, 8(22): 281– 302.

[41] Chang, F., Ilina, O., Lienkamp, M., and L. Voss, M., 'Improving the overall efficiency of automotive inverters using a multilevel converter composed of low voltage Si MOSFETS'. *IEEE Transactions on Power Electronics*, 2019, 34(4): 3586–3602.

[42] R. Teichmann and S. Bernet., 'A comparison of three-level converters versus two-level converters for low-voltage drives, traction, and utility applications'. *IEEE Transactions on Industrial Application*, 2005, 41(3): 855–865.

Chapter 2

Sliding mode control of bidirectional DC–DC converter for EVs

Kanthi Mathew K.[1] *and Dolly Mary Abraham*[1]

Bidirectional DC–DC converter plays a critical role in the development of smart grid applications using electric vehicles. With bidirectional DC–DC converters, the electric vehicle battery returns a portion of the stored energy to the power grid to realize the vehicle-to-grid (V2G) operation. To enhance the overall performance of the bidirectional DC–DC converter, sliding mode controller (SMC) is adopted. The optimal selection of sliding mode parameters is crucial for improving performance and extending the battery life of electric vehicles, which is executed using the Harris Hawks Optimization (HHO) algorithm. The performance of the converter with the SMC is evaluated for different operating conditions during charging and discharging operations. The efficacy of the HHO-based sliding mode controller in the DC–DC converter is verified by comparing it with conventional PI controllers. The results of the tests show that HHO-based sliding mode controller is effective in controlling the charging–discharging properties of bidirectional converters.

2.1 Introduction

Onboard charging technology is becoming more prominent in industry and academia due to its flexibility in charging, promoting the adoption of electric vehicles. Performance enhancement of bidirectional DC–DC converter in the onboard charger is gaining attention in grid-to-vehicle (G2V) and vehicle-to-grid (V2G) operations. Owing to the bidirectional DC–DC converter's high non-linearity and non-minimum phase behavior, a robust controller is required. The sliding mode controller (SMC) is an effective solution for bidirectional converters due to its tracking precision and reliable performance [1–3].

Lithium-ion (Li-ion) batteries have recently achieved popularity in electric vehicle storage applications due to their excellent properties [4,5]. To properly

[1]Department of Electrical Engineering, Government Engineering College, Rajiv Gandhi Institute of Technology, APJ Abdul Kalam Technological University, India

exploit the Li-ion battery's advantages, stable charging circuitry is required. One of the most popular charging methods for Li-ion batteries is constant-current constant-voltage (CC–CV) [6]. The wide range of battery pack equivalent impedance creates the biggest difficulty during the entire charging profile. Therefore, to perform CC–CV charging, precise voltage and current regulation are needed on the battery side. This requirement can be met with an optimal controller design for the bidirectional DC–DC converter.

The linear PI controller is the most popular control technique due to its simplicity [7]. However, it gives poor dynamic performance under a variety of operational circumstances. As bidirectional DC–DC converter exhibits strong non-linearity and non-minimum phase behavior, sophisticated controller design is required. Several non-linear controllers are stated to increase stability, robustness, and voltage regulation. The most common non-linear controllers used with DC–DC converters are model predictive control, fuzzy logic control, neural network control, adaptive control, sliding mode control, and back-stepping control [8–12]. The major limitations of these controllers are their challenging computational and real-time implementation.

Several studies on the application of sliding mode control (SMC) for the drive system and power electronic converters have recently been conducted due to its robustness and ability to minimize the impacts of parameter variation. The study in [13] explained a sliding mode controller for a bidirectional resonant converter with a voltage control topology to improve the dynamics of the system. In [14], PI and SMC are integrated to provide robust performance for both voltage and current regulation of a Cuk converter. In order to provide a fast transient response for the bidirectional DC–DC converter in hybrid energy storage systems, the study in [15] developed an adaptive SMC that requires accurate parameter estimation. A comparative study of hysteresis and pulse width modulation-based SMC for a bidirectional buck–boost converter is presented in [16] to provide a seamless transition between the buck and boost modes of operation. A sliding mode controller based on disturbance observance was designed in [17] to improve the dynamic performance of buck–boost converter. The design of an integral sliding mode controller is described in [18], which is used for bidirectional DC chargers in electric vehicles to regulate the battery current.

The modeling of bidirectional DC–DC converter along with the design phases of the sliding mode controller is presented in this chapter. The sliding mode parameters are selected using HHO algorithm with the aid of the derived stability conditions of the sliding mode controller. With the systematic design procedure for HHO-based SMC provided in this study, major challenges in the practical implementation of SMC are solved.

The converter response is studied for all operating conditions for both charging and discharging modes. The effectiveness of HHO-based sliding mode controller in bidirectional DC–DC converter is demonstrated through the comparison with the conventional PI controllers. The Typhoon hardware in the loop (HIL-402) prototype is used to perform real-time simulation and experimentation to examine the controller's efficacy.

2.2 Sliding mode control of bidirectional DC–DC converter

The schematic circuit of the bidirectional DC–DC converter system is shown in Figure 2.1. The topology includes two power switches S_{b1} and S_{b2}, an inductor L_b, and filter capacitances C_{dc} and C_{bat}. The instantaneous DC bus voltage is represented by v_{dc} and the battery voltage is represented by v_{bat}.

The bidirectional DC–DC converter is regulated as a buck converter for charging purposes by activating switch S_{b1} while maintaining switch S_{b2} inactive and as a boost converter for discharging operations by activating switch S_{b2} while maintaining switch S_{b1} inactive. The control signals for these switches are generated using a sliding mode controller with the aid of the HHO algorithm.

2.2.1 Modeling of the converter

The behavioral model during buck operation in continuous conduction can be written as:

$$\begin{cases} L_b \dfrac{di_{Lb}}{dt} = v_{dc}u - v_{bat} \\ C_{bat} \dfrac{dv_{bat}}{dt} = i_{Lb} - i_{bat} \end{cases} \tag{2.1}$$

where u is the switching state of the switch S_{b1}.

The behavioral model during boost operation in continuous conduction can be written as:

$$\begin{cases} L_b \dfrac{di_{Lb}}{dt} = v_{bat} - v_{dc}\overline{u} \\ C_{dc} \dfrac{dv_{dc}}{dt} = i_{Lb}\overline{u} - i_{dc} \end{cases} \tag{2.2}$$

where u is the switching state of the switch S_{b2} and $\overline{u} = 1 - u$ is the reverse logic of u. The input current and capacitor current at the battery side is indicated by i_{bat} and

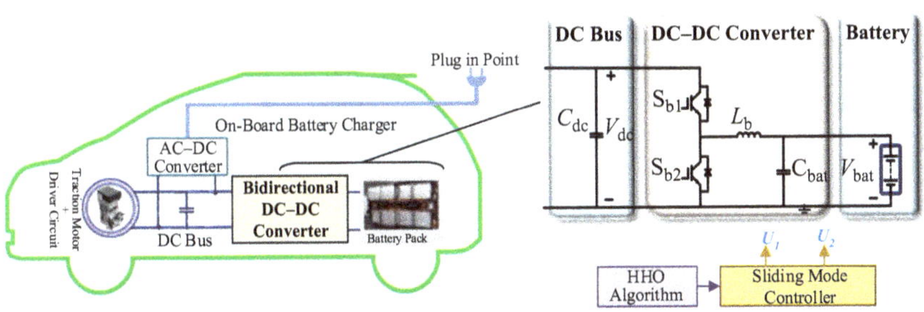

Figure 2.1 Electric vehicle onboard charger architecture and circuit configuration of bidirectional DC–DC converter

i_{Cbat}, while i_{dc} and i_{cdc} indicates the input current and the capacitor current at the DC bus side respectively. i_{Lb} indicates the inductor current.

2.2.2 Choice of sliding surface

Initially, a certain sliding surface is chosen to start the design of the sliding mode controller, and from there a control law is established to bring the state trajectory to the required origin. Finally, the controller's existence criteria are deduced [19]. The inductor's current error x_1, output voltage error x_2, and integral of inductor current and output voltage errors x_3 were chosen as the state variables for this analysis. The sliding surface S is developed by linearly merging the three control variables and is represented as:

$$S = \alpha_1 x_1 + \alpha_2 x_2 + \alpha_3 x_3 \tag{2.3}$$

where α_1, α_2, and α_3 are sliding coefficients.

2.2.3 Derivation of control law

The control equations for SMC are formulated based on the pulse width modulation (PWM) technique. The equivalent control signal, u_{eq}, is first computed by taking the derivative of sliding surface S in (2.3) and assigning $\dot{S} = 0$. The equivalent control signal, u_{eq}, is then correlated with the duty cycle, d, of the PWM technique to obtain a pulse of constant frequency.

2.2.3.1 Charging mode of operation

The control parameters for the bidirectional converter during charging mode are as follows:

$$\begin{cases} x_1 = i_{ref} - i_{Lb} \\ x_2 = V_{batref} - \beta_c v_{bat} \\ x_3 = \int [x_1 + x_2] dt = \int \left(i_{ref} - i_{Lb}\right) dt + \int \left(V_{batref} - \beta_c v_{bat}\right) dt \end{cases} \tag{2.4}$$

where V_{batref} is the battery voltage reference, β_c indicates the feedback network ratio, and i_{ref} is the instantaneous reference value of the inductor current. The battery voltage error is utilized to produce the i_{ref} profile and is given by: $i_{ref} = K_c\ [V_{batref} - \beta_c v_{bat}]$, where K_c is the amplified gain of the battery output voltage error. With the time differentiation of (2.4), the dynamic model of the system is obtained as:

$$\begin{cases} \dot{x_1} = -\dfrac{\beta_c K_c}{C_{bat}} i_{cbat} - \dfrac{(v_{dc} u - v_{bat})}{L_b} \\ \dot{x_2} = -\dfrac{\beta_c}{C_{bat}} i_{cbat} \\ \dot{x_3} = (K_c + 1)\left[V_{batref} - \beta_c v_{bat}\right] - i_{Lb} \end{cases} \tag{2.5}$$

Derivative of the state variable trajectory, $\dot{S}$, is obtained with the substitution of (2.5) as:

$$\begin{aligned}\dot{S} &= \alpha_1\dot{x_1} + \alpha_2\dot{x_2} + \alpha_3\dot{x_3} \\ &= -\frac{\beta_c}{C_{bat}}(\alpha_1 K_c + \alpha_2)i_{cbat} - \alpha_1\frac{v_{dc}}{L_b}u + \alpha_1\frac{v_{bat}}{L_b} \\ &\quad + \alpha_3(K_c \quad + 1)\left(V_{batref} - \beta_c v_{bat}\right) - \alpha_3 i_{Lb}\end{aligned} \tag{2.6}$$

The equivalent control function, u_{eq}, is computed by setting $\dot{S}= 0$ in (2.6) and on rearranging as:

$$\begin{aligned}u_{eq} &= -\frac{\beta_c L_b}{C_{bat}}\left(K_c + \frac{\alpha_2}{\alpha_1}\right)i_{cbat} + \frac{v_{bat}}{v_{dc}} + \frac{\alpha_3}{\alpha_1}\frac{L_b}{v_{dc}}i_{Lb} \\ &\quad + \frac{\alpha_3}{\alpha_1}\frac{L_b(K_c \quad + 1)}{v_{dc}}\left(V_{batref} - \beta_c v_{bat}\right) \\ &= \frac{-K_{c2}i_{Cbat} + v_{bat} + K_{c1}\left(V_{batref} - \beta_d v_{bat}\right) - K_{c3}i_{Lb}}{v_{dc}}\end{aligned} \tag{2.7}$$

where K_{c1}, K_{c2}, and K_{c3} are the sliding mode parameters given by:

$$K_{c1} = \frac{\alpha_3}{\alpha_1}L_b(K_c + 1);\ K_{c2} = \frac{\beta_c L_b}{C_{bat}}\left(K_c + \frac{\alpha_2}{\alpha_1}\right);\ K_{c3} = \frac{\alpha_3}{\alpha_1}L_b \tag{2.8}$$

Substituting (2.7) into the inequality $0 < u_{eq} < 1$ gives

$$0 < -K_{c2}i_{Cbat} + v_{bat} + K_{c1}\left(V_{batref} - \beta_c v_{bat}\right) - K_{c3}i_{lb} < v_{dc} \tag{2.9}$$

The duty ratio control, $0 < d = \frac{V_{control}}{V_{ramp}} < 1$, and (2.9) are correlated to generate the following relationship:

$$\begin{cases} v_c = G_c K_{c1}\left(V_{batref} - \beta_c v_{bat}\right) - G_c K_{c2} i_{Cbat} - G_c K_{c3} i_{Lb} + G_c v_{bat} \\ v_{ramp} = G_c v_{dc} \end{cases} \tag{2.10}$$

where G_c is the scaling factor. The control structure during the charging operation is shown in Figure 2.2.

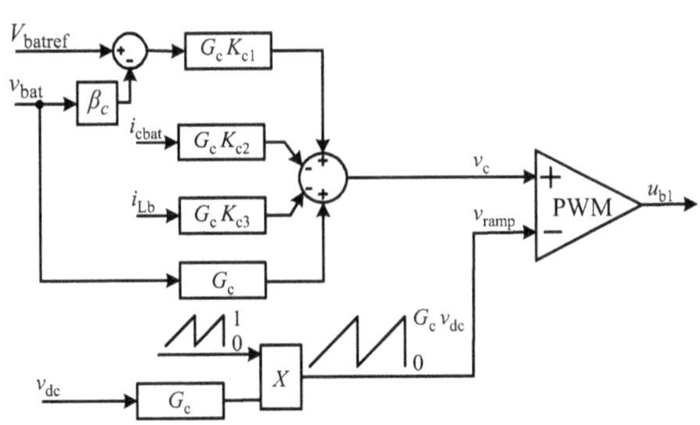

Figure 2.2 Structure of SMC during charging mode

2.2.3.2 Discharging mode of operation

The control parameters for the bidirectional converter during discharging mode are as follows:

$$\begin{cases} x_1 = i_{ref} - i_{Lb} \\ x_2 = V_{dcref} - \beta_d v_{dc} \\ x_3 = \int [x_1 + x_2] dt = \int \left(i_{lref} - i_{Lb}\right) dt + \int \left(V_{dcref} - \beta_d v_{dc}\right) dt \end{cases} \tag{2.11}$$

where V_{dcref} is the DC bus voltage reference, β_d indicates the feedback network ratio, and i_{ref} is the instantaneous reference value of the inductor current. The DC bus voltage error is utilized to produce the i_{ref} profile and is given by: $i_{ref} = K_d [V_{dcref} - \beta_d v_{dc}]$, where K_d is the amplified gain of the DC bus voltage error. With the time differentiation of (2.11), the dynamic model of the system is obtained as:

$$\begin{cases} \dot{x_1} = -\dfrac{\beta_d K_d}{C_{dc}} i_{cdc} - \dfrac{(v_{bat} - v_{dc}\overline{u})}{L_b} \quad \dot{x_2} = -\dfrac{\beta_d}{C_{dc}} i_{cdc} \\ \dot{x_3} = (K_d + 1)\left[V_{dcref} - \beta_d v_{dc}\right] - i_{Lb} \end{cases} \tag{2.12}$$

Derivative of the state variable trajectory, $\dot{S}$, is obtained with the substitution of (2.12) as:

$$\begin{aligned} \dot{S} &= \alpha_1 \dot{x_1} + \alpha_2 \dot{x_2} + \alpha_3 \dot{x_3} \\ &= -\frac{\beta_d}{C_{dc}} (\alpha_1 K_d + \alpha_2) i_{cdc} - \alpha_1 \frac{v_{bat}}{L_b} + \alpha_1 \frac{v_{dc}}{L_b} \overline{u} \\ &+ \alpha_3 (K_d \quad + 1)\left(V_{dcref} - \beta_d v_{dc}\right) - \alpha_3 i_{Lb} \end{aligned} \tag{2.13}$$

The equivalent control function, $\overline{u}_{eq}$, is computed by setting $\dot{S} = 0$ in (2.13) and on rearranging as:

$$\begin{aligned} \overline{u}_{eq} &= \frac{\beta_d L_1}{C_{dc}} \left(K_d + \frac{\alpha_2}{\alpha_1}\right) i_{cdc} + \frac{v_{bat}}{v_{dc}} + \frac{\alpha_3}{\alpha_1} \frac{L_b}{v_{dc}} i_{Lb} \\ &- \frac{\alpha_3}{\alpha_1} \frac{L_b (K_d + 1)}{v_{dc}} \left(V_{dcref} - \beta_d v_{dc}\right) \\ &= \frac{K_{d2} i_{Cdc} + v_{bat} - K_{d1}\left(V_{dcref} - \beta_d v_{dc}\right) + K_{d3} i_{Lb}}{v_{dc}} \end{aligned} \tag{2.14}$$

where K_{d1}, K_{d2}, and K_{d3} are sliding mode parameters given by:

$$K_{d1} = \frac{\alpha_3}{\alpha_1} L_b (K_d + 1); \quad K_{d2} = \frac{\beta_d L_b}{C_{dc}} \left(K_d + \frac{\alpha_2}{\alpha_1}\right); \quad K_{d3} = \frac{\alpha_3}{\alpha_1} L_b \tag{2.15}$$

Since $\overline{u}_{eq}$ is the inverse logic of u_{eq},

$$u_{eq} = 1 - \overline{u}_{eq} = \frac{v_{dc} - K_{d2} i_{Cdc} - v_{bat} + K_{d1}\left(V_{dcref} - \beta_d v_{dc}\right) - K_{d3} i_{Lb}}{v_{dc}} \tag{2.16}$$

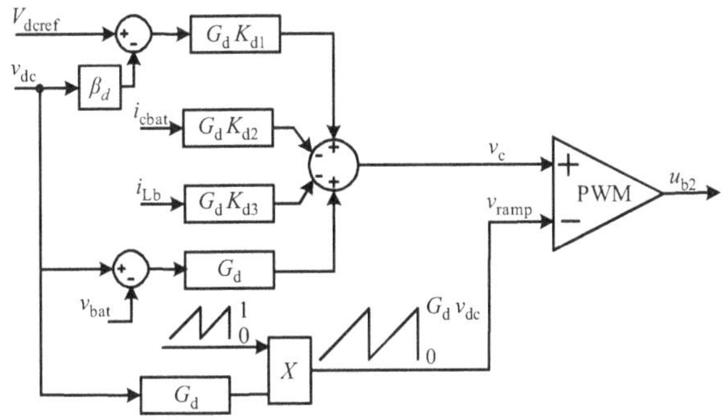

Figure 2.3 Structure of SMC during discharging mode

Substituting (2.16) into the inequality $0 < u_{eq} < 1$ gives

$$0 < v_{dc} - K_{d2} i_{Cdc} - v_{bat} + K_{d1}\left(V_{dcref} - \beta_d v_{dc}\right) - K_{d3} i_{Lb} < v_{dc} \tag{2.17}$$

The duty ratio control, $0 < d = \frac{V_{control}}{V_{ramp}} < 1$, and (2.17) are correlated to generate the following relationship:

$$\begin{cases} v_c = G_d K_{d1}\left(V_{dcref} - \beta_d v_{dc}\right) - G_d K_{d2} i_{Cdc} - G_d K_{d3} i_{Lb} \\ \qquad + G_d\left(v_{dc} - v_{bat}\right) \\ v_{ramp} = G_d v_{dc} \end{cases} \tag{2.18}$$

where G_d is the scaling factor. The control structure during discharging operation is shown in Figure 2.3.

2.2.4 Derivation of existence and stability conditions

The Lyapunov approach is adopted for the stability analysis of the generated sliding mode control law for both operating modes [20]. The finite-time convergence to the sliding surface is ensured if $\lim_{S\to 0} S \cdot \dot{S}$ that can be represented as $\dot{S}_{S\to 0^+} < 0$ and $\dot{S}_{S\to 0^-} > 0$.

2.2.4.1 Charging mode of operation

The first constraint of reachability can be obtained by substitution of $u = 1$ in (2.6) as:

$$\begin{aligned} & -\frac{\beta_c}{C_{bat}}(\alpha_1 K_c + \alpha_2) i_{cbat} - \alpha_1 \frac{v_{dc}}{L_b} + \alpha_1 \frac{v_{bat}}{L_b} \\ & + \alpha_3 (K_c + 1)\left(V_{batref} - \beta_c v_{bat}\right) - \alpha_3 i_{Lb} < 0 \end{aligned} \tag{2.19}$$

The second constraint of reachability can be obtained by substitution of $u = 0$ in (2.6) as:

$$\begin{aligned} & -\frac{\beta_c}{C_{bat}}(\alpha_1 K_c + \alpha_2) i_{cbat} - \alpha_1 \frac{v_{dc}}{L_b} \\ & + \alpha_3 (K_c + 1)\left(V_{batref} - \beta_c v_{bat}\right) - \alpha_3 i_{Lb} < 0 \end{aligned} \tag{2.20}$$

The aforementioned reachability constraints are summarized as follows by applying the steady-state criteria and taking into account (2.8):

$$\begin{cases} -K_{c2}i_{cbat(\min)} + K_{c1}\left(V_{batref} - \beta_c v_{bat(SS)}\right) - K_{c3}i_{Lb(\max)} \\ < v_{dc(\min)} - v_{bat(SS)} \\ K_{c2}i_{cbat(\max)} - K_{c1}\left(V_{batref} - \beta_c v_{bat(SS)}\right) - K_{c3}i_{Lb(\min)} < v_{bat(SS)} \end{cases} \tag{2.21}$$

where $v_{dc(\min)}$ represents the minimum value of DC bus voltages; $v_{bat(SS)}$ indicates the desired steady-state battery voltage which is a DC variable with a small variation from the reference voltage, V_{batref}; whereas $i_{cbat(\min)}$ and $i_{cbat(\max)}$ are the minimum and maximum values of battery side capacitor current respectively when the converter is working at full load conditions.

2.2.4.2 Discharging mode of operation

The first constraint of reachability can be obtained by substitution of $\overline{u} = 1$ in (2.13) as:

$$\begin{aligned} &-\frac{\beta_d}{C_{dc}}(\alpha_1 K_d + \alpha_2)i_{cdc} - \alpha_1\frac{v_{bat}}{L_b} \\ &+ \alpha_3(K_d + 1)\left(V_{dcref} - \beta_d v_{dc}\right) - \alpha_3 i_{Lb} > 0 \end{aligned} \tag{2.22}$$

The second constraint of reachability can be obtained by substitution of $\overline{u} = 0$ in (2.13) as:

$$\begin{aligned} &-\frac{\beta_d}{C_{dc}}(\alpha_1 K_d + \alpha_2)i_{cdc} - \alpha_1\frac{v_{bat}}{L_b} + \alpha_1\frac{v_{dc}}{L_b} \\ &+ \alpha_3(K_d + 1)\left(V_{dcref} - \beta_d v_{dc}\right) - \alpha_3 i_{Lb} > 0 \end{aligned} \tag{2.23}$$

The aforementioned reachability constraints are summarized as follows by applying the steady-state criteria and taking them into account (2.15):

$$\begin{cases} -K_{d2}i_{cdc(\min)} + K_{d1}\left(V_{dcref} - \beta_d v_{dc(SS)}\right) - K_{d3}i_{Lb(\max)} < v_{bat(\min)} \\ K_{d2}i_{cdc(\max)} - K_{d1}\left(V_{dcref} - \beta_d v_{dc(SS)}\right) + K_{d3}i_{Lb(\min)} \\ < v_{dc(SS)} - v_{bat(\max)} \end{cases} \tag{2.24}$$

where $v_{bat(\min)}$ and $v_{bat(\max)}$ represents the minimum and maximum values of battery voltages respectively; $v_{dc(SS)}$ indicates the desired steady-state DC bus voltage which is a DC variable with a small variation from the reference voltage, V_{dcref}; whereas $i_{cdc(\min)}$ and $i_{cdc(\max)}$ are the minimum and maximum values of DC bus side capacitor current respectively when the converter is working at full load conditions.

2.2.5 Sliding mode parameter selection using HHO algorithm

The controller performance is significantly governed by the selection of sliding mode parameters. Metaheuristics optimization algorithms are well suited for

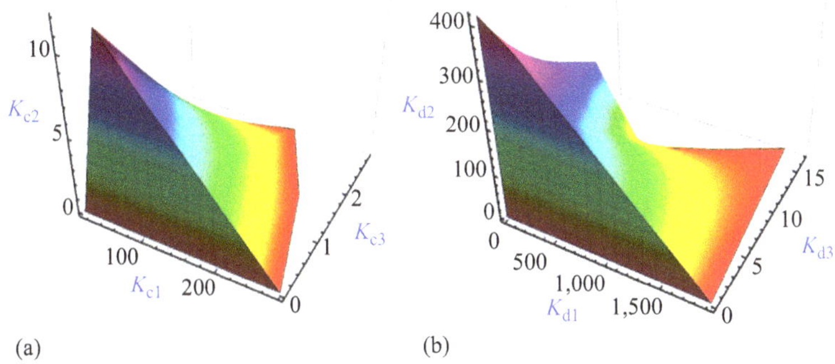

Figure 2.4 Searching range of SMC parameters. (a) Charging mode. (b) Discharging mode.

parameter optimization because of their capacity to resolve challenging optimization issues driven by model uncertainties [21]. Recently, HHO is gaining popularity as it offers faster convergence for solving practical challenges [22–24]. In this work, the HHO algorithm is chosen for selecting SMC parameters. The lowest and highest ranges for SMC parameters are figured from the stability conditions of the controller. The range for the parameters, K_{c1}, K_{c2}, and K_{c3} for charging mode is obtained from the inequalities given in (2.21). In discharging mode, the range for the parameters, K_{d1}, K_{d2}, and K_{d3} is obtained from the inequalities given in (2.24). Using the computational software Mathematica (Wolfram, Version 12.0), the three-dimensional area specified by the constraints is shown in Figure 2.4.

The HHO algorithm's primary goal is to reduce the objective function for improved controller parameter values. The four objective functions that are most frequently used are the integral of squared error (ISE), integral absolute error (IAE), integral of time-weighted squared error (ITSE), and integral of time-weighted absolute error (ITAE). The performance objective, which is written as: $O = \int_0^T t|e(t)|dt$, produces the best results [25,26], and T indicates when the response has reached its steady-state. The system's errors, which comprise the inductor current error, $\left(i_{ref} - i_{L1}\right)$, the battery voltage error, $\left(V_{batref} - \beta_c v_{bat}\right)$ and the DC bus voltage error, $\left(V_{dcref} - \beta_d v_{dc}\right)$, are added up to form the error $e(t)$ in the objective function. Each time the HHO algorithm executes, the objective function is evaluated. When the phase is complete, the best value is applied to the structure of SMC.

The performance of the optimization algorithm is based on two stages: exploration and exploitation. The HHO algorithm is a powerful search algorithm maintaining a fine proportion between these two stages. In HHO, the Harris Hawks indicate the optimal results and the prey indicates the best result in each iteration.

2.2.5.1 Exploration stage

In this stage, the position of Hawks is updated using two strategies. First, the Hawks randomly update the location such as choosing a random tall tree to locate the prey. Second, the update is dependent on the location of the prey and other family members. The mathematical expression of these two strategies is as follows:

$$x^{m+1} = \begin{cases} x_r^m - r_1 \left| x_r^m - 2r_2 x^m \right|, q \geq 0.5 \\ \left(x_{bst}^m - x_{avg}^m \right) - r_3 (l_l + r_4 (u_l - l_l)), q < 0.5 \end{cases}$$

$$x_{avg}^m = \frac{1}{n} \sum_{i=1}^{n} x_i^m \tag{2.25}$$

where x is the position of the Hawks, x_{bst} is the position of the prey, x_r is the random position of the Hawks, l_l is the lower limit, u_l is the upper limit of the parameters, r_1, r_2, r_3, r_4, and q are random numbers in the range (0,1), m is the number of iterations, and x_{avg} is the average position of the Hawks.

2.2.5.2 Migration from the exploration stage to the exploitation stage

The stages of HHO are changed depending on the escaping energy (E) of the prey, which is decreasing with the iteration count. This is represented as follows:

$$E = 2E_0 \left(1 - \frac{m}{m_{\max}} \right), E_0 = 2r_e - 1 \tag{2.26}$$

where $m_{\max}$ is the maximum iteration count, E_0 is the initial escaping energy randomly varying between (−1,1), and r_e is a random number in the range (0,1). When $E \geq 1$, HHO executes the exploration stage, and when $E < 1$, the algorithm shifts to the exploitation stage.

2.2.5.3 Exploitation stage

In this stage, the Hawks chase the prey using four different strategies depending on the remaining escaping energy and the escaping probability (r) of the prey. When $r < 0.5$, the prey can exit, and when $r \geq 0.5$, the prey fails to exit before the surprise attack of the Hawks. To trap the prey, the Hawks will engage in a soft besiege when $E \geq 0.5$ or a hard besiege when $E < 0.5$. If these two strategies fail, the Hawks imitate the escape movement of the prey by rapidly diving and adjusting their flight direction and location.

Soft besiege ($r \geq 0.5$ and $E \geq 0.5$): This strategy is modeled by the following expressions:

$$x^{m+1} = \Delta x^m - E \left| J x_{bst}^m - x^m \right|, \quad \Delta x^m = x_{bst}^m - x^m, \quad J = 2(1 - r_5) \tag{2.27}$$

where Δx is the difference in location between the prey and the Hawks. The term J is the jump strength of the prey randomly varying in each iteration with $r_5 \in (0, 1)$.

Hard besiege ($r \geq 0.5$ and $E < 0.5$): In this strategy, the position of the Hawks is updated using:

$$x^{m+1} = x^m_{bst} - E|\Delta x^m| \tag{2.28}$$

Soft besiege with progressive rapid dives ($r < 0.5$ and $E \geq 0.5$): This stage utilizes the concept of Levy Flight (LF) to simulate the natural zigzag movement of the prey. This scenario is implemented by the flowing rules:

$$\begin{aligned} x^{m+1} &= \begin{cases} Y, & if \quad F(Y) < F(x^m) \\ Z, & if \quad F(Z) < F(x^m) \end{cases} \\ Y &= x^m_{bst} - E|Jx^m_{bst} - x^m|, Z = Y + S \times LF(D) \end{aligned} \tag{2.29}$$

where Y is the current movement, Z is the sudden dive of the hawks, D is the parameter dimension, $S \in R^{1\times D}$ is a random vector, and LF function is computed using:

$$\begin{aligned} LF(D) &= 0.01 \times \frac{u \times \sigma}{|v|^{\frac{1}{\beta}}}, u \in (0,1), \sigma = \left(\frac{\Gamma(1+\beta) \times \sin\left(\frac{\pi\beta}{2}\right)}{\Gamma\left(\frac{(1+\beta)}{2}\right) \times \beta \times 2^{\left(\frac{(\beta-1)}{2}\right)}} \right)^{\frac{1}{\beta}} \\ \beta &= constant = 1.5 \end{aligned} \tag{2.30}$$

Hard Besiege with progressive rapid dives ($r < 0.5$ and $E < 0.5$): In this stage, the Hawks aim to decrease the distance between their average locations. This strategy is formulated as follows:

$$\begin{aligned} x^{m+1} &= \begin{cases} Y', & if \quad F(Y') < F(x^m) \\ Z', & if \quad F(Z') < F(x^m) \end{cases} \\ Y' &= x^m_{bst} - E\left|Jx^m_{bst} - x^m_{avg}\right|, Z' = Y' + S \times LF(D) \end{aligned} \tag{2.31}$$

Figure 2.5 shows the flowchart for the optimization procedure, and Table 2.1 shows the HHO algorithm's input data.

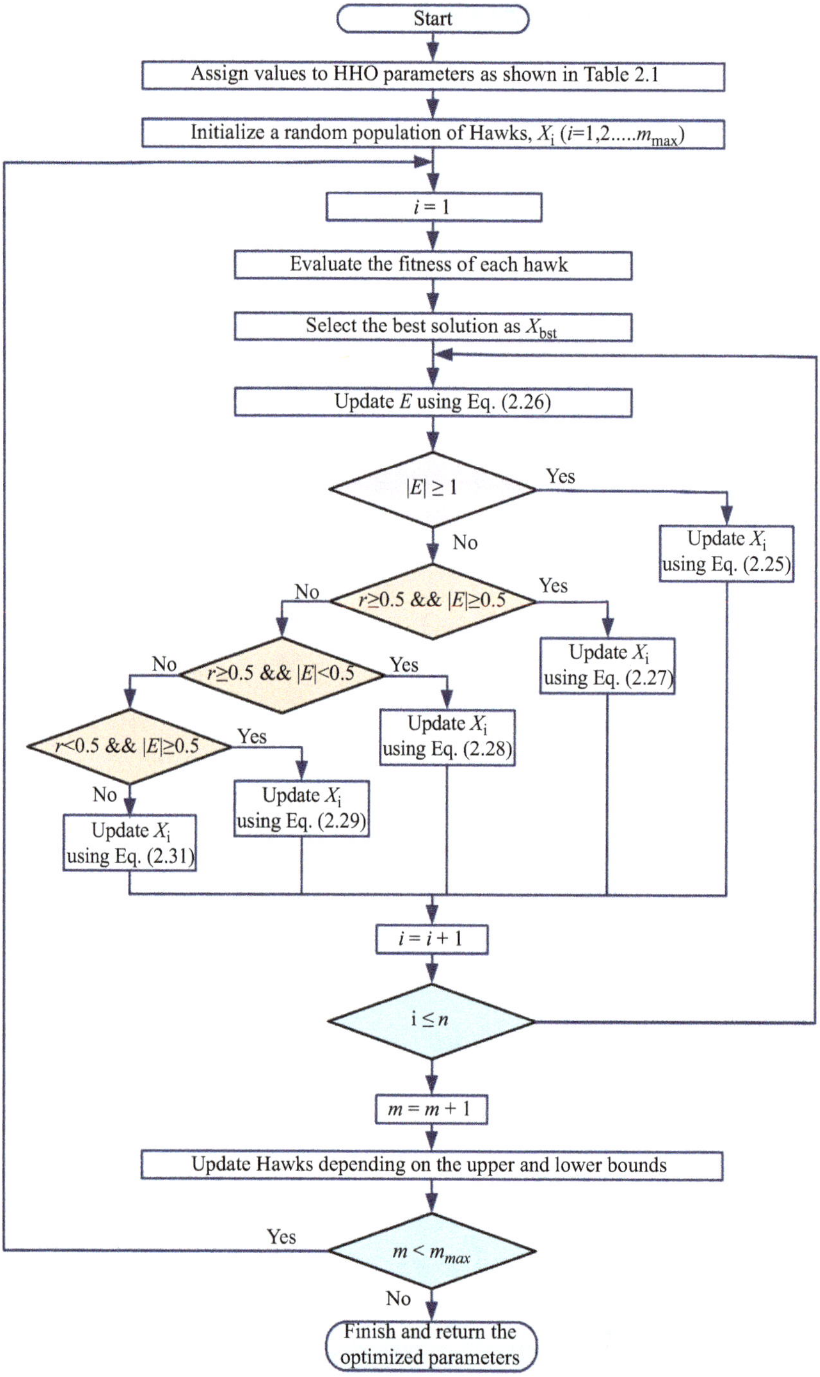

Figure 2.5 Flowchart of optimization process

Table 2.1 Variables of HHO algorithm

Descriptions		Variable	Value
Population size		n	20
Iteration count		$m_{\max}$	50
Dimension		D	3
Charging mode	Upper limit	u_l	[300 12 3]
$[K_{c1} \quad K_{c2} \quad K_{c3}]$	Lower limit	l_l	[0 0 0]
Discharging mode	Upper limit	u_l	[2,000 400 15]
$[K_{d1} \quad K_{d2} \quad K_{d3}]$	Lower limit	l_l	[0 0 0]

2.3 Simulation and experimental verifications

A test setup with the Typhoon Hardware-in-the-Loop device (HIL-402) was developed, as shown in Figure 2.6, to evaluate the performance of the bidirectional DC–DC converter with sliding mode controller. The results were recorded with the LeCroy Wavesurfer 44 Xs. Table 2.2 shows the specifications of the bidirectional DC–DC converter. The SMC parameters for both modes of operation were optimized using the HHO algorithm, as shown in Table 2.3. The HHO algorithm was executed in MATLAB® (R2019b) software.

In this study, a comparison of HHO-SMC versus PI controller for the bidirectional DC–DC converter was provided. The steady-state performance during the charging mode of operation was carried out in HIL SCADA using the Typhoon HIL control center (version 2020.2) as shown in Figure 2.7. The CC–CV profile was used to charge the Li-ion battery. The battery is charged in CC mode at a constant current rate of 0.2C rate (30 A in this work), while the battery voltage gradually increases. The CV mode activates when the battery voltage hits the study's predetermined level of 109 V. In the CV mode of operation, the battery voltage is maintained constant, and the battery current gradually falls. Figure 2.7(a) shows the CC–CV charging with a PI controller, which exhibits switching transients during the transition from CC to CV mode. It was observed from Figure 2.7(b) that the switching transients during the mode change from CC to CV are mitigated with a sliding mode controller.

Figure 2.8 depicts the performance comparison between the cascaded PI and HHO-based SMC during transient analysis. The transient response of the system in charging mode was evaluated with a deviation of $\pm$20 V from the nominal value of the input DC bus voltage (370 V). Furthermore, the battery current was alternated with three different values (i_{bat} = 30 A, 20 A, and 10 A) for the analysis. From Table 2.4, it is clear that the SMC offers minimal ripple content in battery current (less than 5%) compared to the PI controller under different operating conditions.

Figure 2.9 displays the control voltage, ramp voltage, and gate pulse steady-state waveforms of SMC in the charging and discharging modes. It was evident that

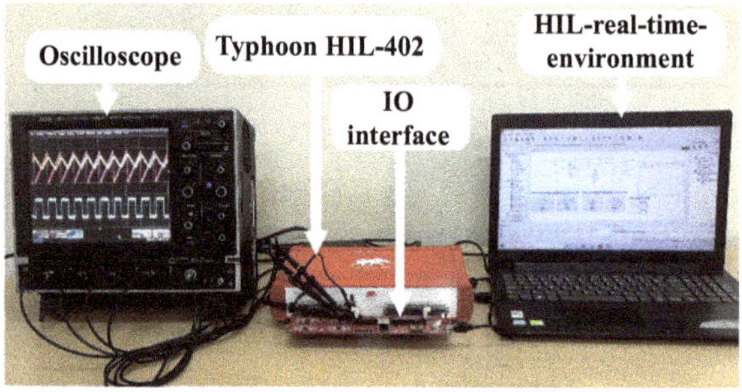

Figure 2.6 Experimental setup of hardware in the loop using HIL-402 system

Table 2.2 Specifications of DC–DC converter

Descriptions	**Parameters**	**Nominal values**
Maximum power	P_{max}	3.3 kW
Switching frequency	f_{sw}	20 kHz
DC link voltage	V_{dc}	370 V
Battery voltage	V_{bat}	96 V
Battery	Li-ion	150 Ah
Capacitances	C_{dc}, C_{bat}	10 mF, 5 mF
Inductances	L_b	2.17 mH

Table 2.3 Parameters of the sliding mode controller

Mode of operation	**Parameters**	**Nominal values**
Charging mode	$G_c = \beta_c$	1/18
Parameters	K_{c1}, K_{c2}, K_{c3}	18.7, 0.25, 0.95
Discharging mode	$G_d = \beta_d$	1/37
Parameters	K_{d1}, K_{d2}, K_{d3}	409.1, 7.6, 5.24

the frequency was maintained at 20 kHz under all conditions. As a result, there are a fewer EMI issues, which is advantageous for DC–DC converters in the onboard charger.

The converter's discharging mode performance was evaluated for a range of load variations. Table 2.5 provides the fluctuation in steady-state DC bus voltage for various load conditions. SMC gives a 10% improvement in full load conditions

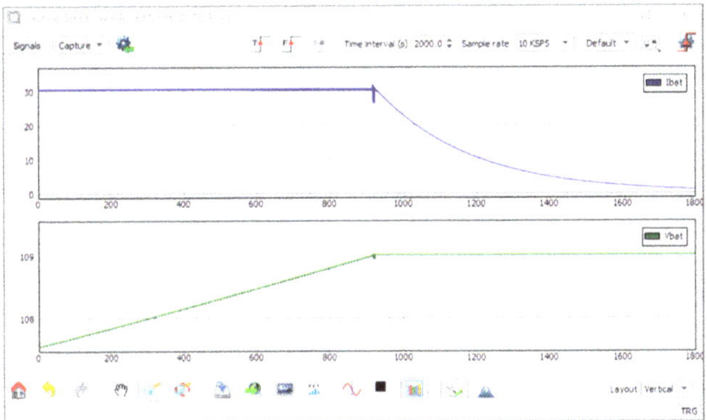

(a)

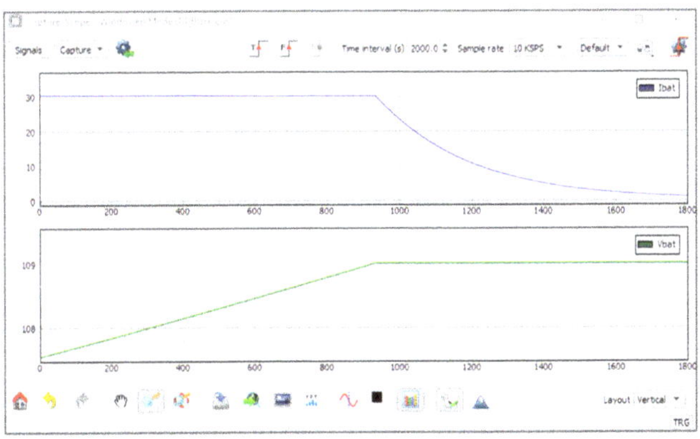

(b)

Figure 2.7 Real-time simulation of CC–CV charging of the battery. (a) PI. (b) SMC.

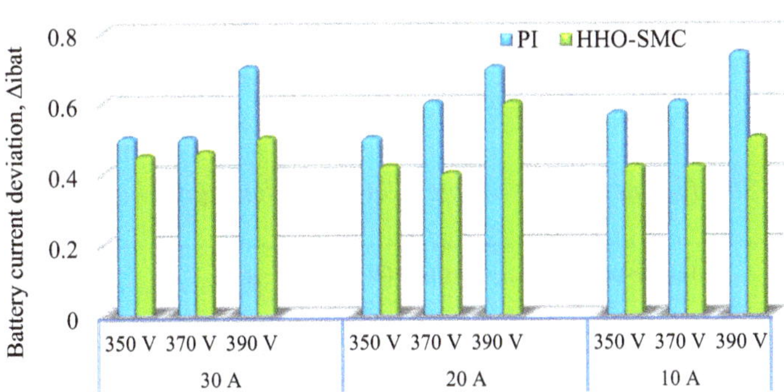

Figure 2.8 Battery current deviation, Δi_{bat} for different input DC bus voltage and battery charging current

Table 2.4 Battery current regulation property for different charging current and DC bus voltage

Input DC bus voltage	Percentage change in battery current $= \left(\frac{\Delta i_{bat}}{i_{bat}}\right) \times 100$ (%)					
	$i_{bat} = 30$ A		$i_{bat} = 20$ A		$i_{bat} = 10$ A	
	PI	SMC	PI	SMC	PI	SMC
350 V	1.7	1.5	2.5	2.1	5.7	4.2
370 V	1.7	1.5	3	2	6	4.2
390 V	2.3	1.7	3.5	3	7.4	5

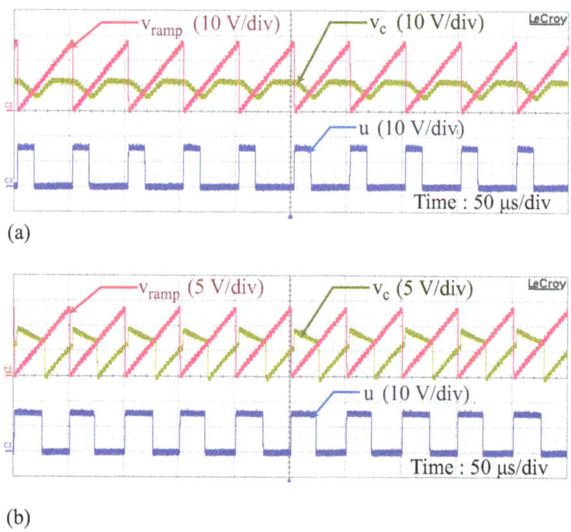

Figure 2.9 Control voltage, ramp voltage, and gate pulse ($V_{control}$, V_{ramp}, and u) of HHO–SMC during charging and discharging mode. (a) Charging mode. (b) Discharging mode.

Table 2.5 Steady-state DC bus voltage deviation during discharging mode

Power	Steady-state voltage ripple (V)		Performance enhancement (%)
	PI	SMC	
3.3 kW	0.39	0.35	10.3
1.67 kW	0.25	0.21	16
330 W	0.23	0.22	4.4

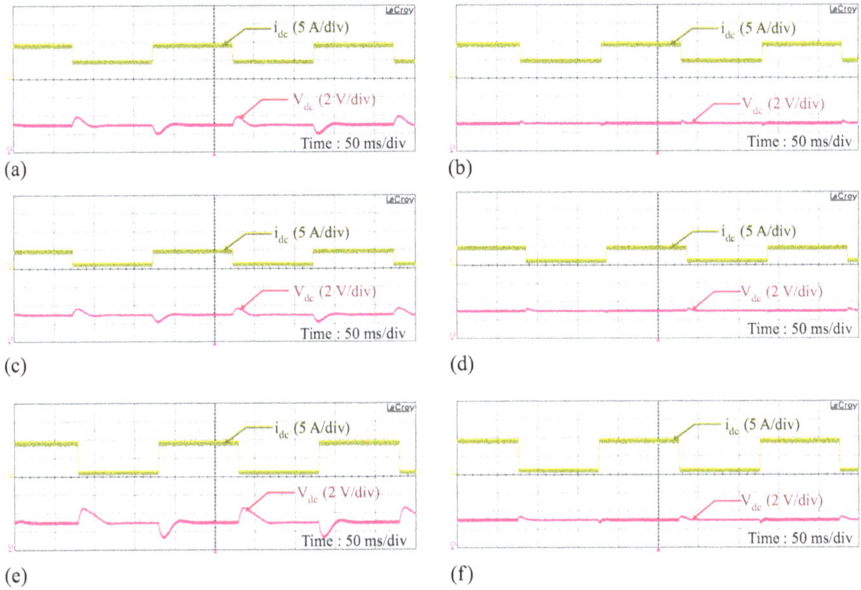

Figure 2.10 Experimental results of DC bus voltage, and current (v_{dc}, i_{dc}) alternating between full (8.9 A), half (4.5 A), and minimum (0.89 A) load. (a) PI (8.9/4.5 A), (b) SMC (8.9/4.5 A), (c) PI (4.5/0.89 A), (d) SMC (4.5/0.89 A), (e) PI (8.9/0.89 A), (f) SMC (8.9/0.89 A).

while maintaining almost the same ripple content in DC bus voltage under varied load changes.

To examine the transient performance of the DC–DC converter during discharge mode, the analysis was performed by switching the output DC bus current between i_{dc} = 8.9 A (full load), i_{dc} = 4.5 A (half load), and i_{dc} = 0.89 A (minimum load). Figure 2.10 shows the converter's transient performance with the two controllers. It has been noted that when the load is changed from maximum to half or minimum value, the DC bus output voltage undershoots, and when the load is changed from half or minimum value to maximum, it overshoots. For various operating conditions, the PI controller displays high transient behavior and a large settling time. However, the dynamic behavior of SMC is consistent with a smaller ripple and shorter settling time.

Table 2.6 shows the overview of the converter's load regulation properties in the discharging mode with the cascaded PI and sliding mode controller. The sliding mode controller reduces overshoot by an average of 65.7% and undershoots by an average of 69.2%. Furthermore, the settling time was shortened on average by 60.9%.

Table 2.6 Load regulation property during discharging mode

Scenario	3.3 kW → 1.67 kW			1.67 kW → 330 W			3.3 kW → 330 W		
	PI	SMC	Performance enhancement	PI	SMC	Performance enhancement	PI	SMC	Performance enhancement
Overshoot (V)	1.68	0.66	60.9%	1.32	0.54	58.9%	2.67	0.6	77.4%
Undershoot (V)	1.8	0.81	55%	1.46	0.38	74.1%	2.98	0.64	78.4%
Settling time (ms)	36	14	61.1%	25	13	48%	38	10	73.7%

2.4 Conclusion

In this chapter, the process of designing sliding mode controllers for bidirectional converter's charging and discharging modes of operation was detailed. The sliding mode parameters were chosen using the HHO algorithm according to the stability criteria of the sliding mode controller. A comparative study was performed for the steady-state and dynamic performance of the bidirectional converter with the sliding mode controller and the PI controller. The results were verified by employing hardware in the loop real-time prototype. According to the results from the real-time simulation, the shift from CC to CV mode with sliding mode control was smooth. The optimal sliding mode controller offers better load regulation features with reference to settling time undershoot and overshoot. The results of the tests show that the sliding mode controller was effective in controlling the charging–discharging properties of the bidirectional converter.

References

[1] Gorji SA, Sahebi HG, Ektesabi M, and Rad AB. Topologies and control schemes of bidirectional DC–DC power converters: an overview. *IEEE Access.* 2019;7:117997–8019.

[2] Huangfu Y, Guo L, Ma R, and Gao F. An advanced robust noise suppression control of bidirectional DC–DC converter for fuel cell electric vehicle. *IEEE Transactions on Transportation Electrification.* 2019;5(4):1268–1278.

[3] Liu S, Liu X, Jiang S, *et al.* Application of an improved STSMC method to the bidirectional DC–DC converter in photovoltaic DC microgrid. *Energies.* 2022;15(5):1636.

[4] Wang Z, Feng G, Zhen D, Gu F, and Ball A. A review on online state of charge and state of health estimation for lithium-ion batteries in electric vehicles. *Energy Reports.* 2021;7:5141–5161.

[5] Makeen P, Ghali HA, and Memon S. A review of various fast charging power and thermal protocols for electric vehicles represented by lithium-ion battery systems. *Future Transportation.* 2022;2(1):281–299.

[6] Gao Y, Zhang X, Cheng Q, Guo B, and Yang J. Classification and review of the charging strategies for commercial lithium-ion batteries. *IEEE Access.* 2019;7:43511–43524.

[7] Mumtaz F, Zaihar Yahaya N, Tanzim Meraj S, Singh B, Kannan R, and Ibrahim O. Review on non-isolated DC-DC converters and their control techniques for renewable energy applications. *Ain Shams Engineering Journal.* 2021;12(4):3747–3763.

[8] Albira ME and Zohdy MA. Adaptive model predictive control for DC–DC power converters with parameters' uncertainties. *IEEE Access.* 2021;9:135121–135131.

[9] Yan W, Wei L, Zhan D, Su D, and Zhang M. Robustness analysis and controller design based on a generalized model of nonisolated multiphase DC–DC converter. *IEEE Access.* 2022;10:14846–14856.

[10] Qi Q, Ghaderi D, and Guerrero JM. Sliding mode controller-based switched-capacitor-based high DC gain and low voltage stress DC–DC boost converter for photovoltaic applications. *International Journal of Electrical Power & Energy Systems*. 2021;125:106496.

[11] Dong W, Li S, Fu X, Li Z, Fairbank M, and Gao Y. Control of a buck DC/DC converter using approximate dynamic programming and artificial neural networks. *IEEE Transactions on Circuits and Systems I: Regular Papers.* 2021;68(4):1760–1768.

[12] Komurcugil H. Sliding mode control strategy with maximized existence region for DC–DC buck converters. *International Transactions on Electrical Energy Systems.* 2020;31(3):12764.

[13] Zou S, Mallik A, Lu J, and Khaligh A. Sliding mode control scheme for a CLLC resonant converter. *IEEE Transactions on Power Electronics.* 2019;34(12):12274–84.

[14] Padmanaban S, Ozsoy E, Fedák V, and Blaabjerg F. Development of sliding mode controller for a modified boost Ćuk converter configuration. *Energies.* 2017;10(10):1513.

[15] Fu Z, Wang Y, Tao F, and Si P. An adaptive nonsingular terminal sliding mode control for bidirectional DC–DC converter in hybrid energy storage systems. *Canadian Journal of Electrical and Computer Engineering.* 2020;43(4):282–289.

[16] Mendez-Diaz F, Pico B, Vidal-Idiarte E, Calvente J, and Giral R. HM/PWM seamless control of a bidirectional buck–boost converter for a photovoltaic application. *IEEE Transactions on Power Electronics.* 2019;34(3):2887–2899.

[17] Su M, Feng W, Jiang T, Guo B, Wang H, and Zheng C. Disturbance observer-based sliding mode control for dynamic performance enhancement and current-sensorless of buck/boost converter. *IET Power Electronics*. 2021;14(8):1421–1432.

[18] Cheddadi Y, Idrissi ZE, Errahimi F, and sbai NE. Robust integral sliding mode controller design of a bidirectional DC charger in PV-EV charging station. *International Journal of Digital Signals and Smart Systems.* 2021;5(2):137–151.

[19] Siew-Chong T, Lai YM, and Tse CK. *Sliding Mode Control of switching Power Converters: Techniques and Implementation*. CRC press, 2018.

[20] Siew-Chong T, Lai YM, and Tse CK. General design issues of sliding-mode controllers in DC–DC converters. *IEEE Transactions on Industrial Electronics.* 2008;55(3):1160–1174.

[21] Singh, Pushpendra, Nand K. Meena, Jin Yang, and Adam Slowik. *Swarm Intelligence Algorithms: A Tutorial*. CRC press, 2020: 1–15.

[22] Heidari AA, Mirjalili S, Faris H, Aljarah I, Mafarja M, and Chen H. Harris hawks optimization: algorithm and applications. *Future Generation Computer Systems.* 2019;97:849–872.

[23] Alabool HM, Alarabiat D, Abualigah L, and Heidari AA. Harris hawks optimization: a comprehensive review of recent variants and applications. *Neural Computing and Applications.* 2021;33(15):8939–8980.

[24] Çetinbaş İ, Tamyürek B, and Demirtaş M. Sizing optimization and design of an autonomous AC microgrid for commercial loads using Harris Hawks Optimization algorithm. *Energy Conversion and Management.* 2021;245: 114562.
[25] Banerjee S, Ghosh A, and Rana N. An improved interleaved boost converter with PSO-based optimal type-III controller. *IEEE Journal of Emerging and Selected Topics in Power Electronics.* 2017;5(1):323–337.
[26] Mathew KK and Abraham DM. Particle swarm optimization based sliding mode controllers for electric vehicle onboard charger. *Computers & Electrical Engineering*. 2021;96:107502.

Chapter 3

High-gain DC–DC converter with extremum seeking control for PV application

Nima Ayobi[1], *Saeid Deliri*[2], *Asaad Seyedrahmani*[1], *Amin Ziaei*[3], *Kazem Varesi*[2] *and Sanjeevikumar Padmanaban*[4]

This chapter presents a novel approximation-based extremum-seeking control (AESC) method for the maximum power point tracking (MPPT) problem for the solar photovoltaic (PV) system. Generally, rapid solar radiation and temperature variations make the MPPT problem more challenging in the solar PV system. Besides solving the MPPT problem, the model-free nature of the suggested (AESC) method makes it easy to implement. This feature leads to more precise and faster tracking of maximum power point (MPP) in sudden changes of ambient temperature and solar irradiation as an important factor in MPPT. A high voltage-gain DC–DC converter is utilized in combination with the AESC approach to track the MPP which acts as a power interface between load and panels. The utilized structure has several advantages such as high gain step-up ability per used devices, continuous input current, simple structure, and common ground between source and load. Simulations were performed in MATLAB®–Simulink® and the proposed approach was studied for two scenarios that investigate the impacts of solar radiation and temperature on the solar system. Based on simulation results, the proposed method reduces the oscillations around the MPP of the PV system more significantly compared to the conventional P&O method and tracks MPP faster than the conventional P&O method.

3.1 Introduction

Nowadays, the world population growth, restriction of fossil fuel sources, and environmental issues like air pollution and global warming (caused mainly by using

[1]Faculty of Electrical Engineering, Sahand University of Technology, Iran
[2]Power Electronics Research Laboratory (PERL), Sahand University of Technology, Iran
[3]Faculty of Electrical and Computer Engineering, University of Tabriz, Iran
[4]University of South-Eastern Norway, Norway

fossil fuels) have led to a worldwide increase in the utilization and implementation of renewable energy resources such as photovoltaic panels (PVs), wind turbines, hydroelectric, and biomass. Among various renewable energy resources, PVs have received more attention because of their accessibility, capability of implementation in both portable and/or static applications, and environmentally friendly nature [1]. In recent years, the development in solid-state technology and photovoltaic-panel manufacturing, as well as the advancement of power electronic converters (including DC–DC converters and inverters) have reduced the implementation and utilization cost of solar energy harvesting, which accordingly has led to a considerable increase in worldwide PV-power generation [2,3]. The worldwide increase in solar energy utilization has highlighted the importance of implementation cost and PV power generation efficiency for micro and macro-PV power plant owners. The PVs form the building block of solar power plants, where the generated power mainly depends on the ambient condition such as irradiation and temperature, while a reduction is unavoidable by the aging of panels. While suffering from low overall efficiency, the PVs have a nonlinear current–voltage (I–V) characteristic curve. The maximum power of PV can be extracted at the knee point of its I–V curve. The knee point, as well as the maximum power, varies by the change of irradiation and ambient temperature. Hence, efficient control methods are required to guarantee the MPPT and extraction of PV's peak power. Many different methodologies have been presented in the literature. The continuously varying nature of ambient temperature and irradiation is the main challenge on the MPPT of PVs, which can be tackled by adopting efficient control methodologies [4]. In other words, in proposed MPPT methods in the literature, the DC–DC converter acts as a controllable power interface between solar cells and load (resistive load or battery) to perform impedance matching between a solar cell and load for maximum power transfer to the load [5]. Various inverter structures can be utilized for power injection produced by solar cells to the primary upstream grid. The utilization of inverters for power injection to the primary grid has several advantages, including active and reactive power control and current harmonic compensation. However, using the ancillary capabilities of inverters may affect the efficiency of MPPT methods [6,7]. Researchers in the literature have proposed various MPPT methods. Conventional P&O is the most popular MPPT method, among others. This method performs MPPT based on the P–V curve of the solar cells. The predetermined perturbation step in the duty cycle is applied to collect gradient information and the direction of perturbation to determine the MPP. Unlike soft computing methods, real-time implementation of the P&O method on typical processors is feasible; the method does not require frequent parameter tuning for various solar panels and converters. It can be implemented on both digital and analogue platforms, so this advantage makes P&O an appropriate method for scientific, commercial, and real-time applications. Meanwhile, the conventional P&O method has three main shortcomings, first, the wrong direction of tracking in fast insolation changes, second, the inability to track MPP in shade conditions and third, power loss due to oscillations around MPP. In this method, significant perturbation steps lead to fast convergence. However, it increases power loss at the MPP, while small perturbation steps lead to slower convergence but less power loss at the MPP. Therefore, in this method, perturbation size and frequency must be appropriately selected [8]. In practical implementations, the appropriate sampling rate ranges in portable applications like electric vehicles and portable power plants differs from static power plants [9]. In [10], the model predictive control approach is utilized for MPPT in which the solar cell current is predicted. This leads to the elimination of the current sensor and reduces the cost of the practical implementation of the proposed system. On the other side, physical constraints can be integrated into the control signal of the converter. However, model predictive control needs an accurate model of the system under study since the nonlinear nature of the system and parameter variations of the components, accurate modelling may not be possible, which leads to reducing the efficiency of the suggested method. Also, the practical implementation of predictive model controllers needs expensive processors due to high computational volume. Shams *et al.* [11] have suggested an improved method based on the butterfly optimization approach for MPPT. A SEPIC converter was used as an energy conversion interface. The convergence rate of the suggested method was high, and the method was capable of detecting and tracking MPP in a variety of operating conditions, including direct radiation, shade, and load variation where the method uses just one dynamic parameter as tuning coefficient and the power efficiency of the method is high. However, despite all of the mentioned advantages, the method requires powerful processors for real-time implementation due to the high processing volume.

The problem of investigating and practically implementing an intelligent particle swarm optimization based on fuzzy logic control was studied in [12], where a buck–boost zeta converter was utilized. The results indicate that the suggested method has high power efficiency and more accurate MPPT in direct and shade conditions, ambient temperature, and load fluctuations. In [13], a complex hybrid method based on the combination of INC, ANFIS, and HSFLA-PS was proposed for MPPT, and the performance of the proposed method was investigated through simulations. Utilization of INC and ANFIS combination reduces the required samples for training besides of high power efficiency of the proposed method. Despite this, integrating several complex computational methods for MPPT leads to difficulty in real-time implementations. Li *et al.* [14] suggested a modified method based on a beta algorithm that can track the global MPP and has proper tracking in shading conditions. The method's efficiency was investigated using buck–boost in both simulation and practical implementation, where the accuracy and convergence speed of the method is high. However, the practical implementation of the proposed method may be difficult in commercial applications due to the algorithm's complexity. The MPPT method based on student psychology optimization was investigated in [15]. The student psychology optimization method is categorized as a population-based algorithm. The suggested method is capable of global MPPT. This method was compared with five different methods using simulations, where the results indicate the high performance of the method. Despite all of this, small changes in ambient temperature reduce the performance of MPPT. In [16], a hybrid method based on P&O and PSO with variable perturbation frequency and zero oscillations were proposed. The performance of the suggested method was investigated using SEPIC converter in both simulations and practical implementation and results show accurate and fast-tracking of the global MPP; however, real-time implementation of the PSO method requires considerable computational resources. In MPPT methods, population-based optimization methods such as PSO have better performance than evolutionary ones such as genetic algorithm [17].

Issaadi *et al.* [18] proposed an MPPT method based on a robust neural network against noise with accurate and fast MPPT capability. The use of a neural network in MPPT is very appropriate due to the nonlinear nature of the system under study; however, in fast-changing conditions of environmental parameters like insolation and temperature, the real-time implementation of the method is challenging, and finding the best initial values for neural network training may not be possible [19]. In this chapter, to tackle the mentioned shortcomings in reviewed papers, an AESC method was investigated for MPPT of PV systems. Extremum-seeking control (ESC) methods are adaptive control-based approaches that have a model-free nature and are suitable for real-time optimization applications. In these methods, adjustable parameters must be tuned so that a nonlinear map's extremum points (minimum or maximum) are detected and then maintained [20]. ESC approaches are suitable for low-cost applications such as optimal energy management in buildings due to their model-free nature. Moreover, unlike model-based methods such as model predictive control, it does not require accurate or exact modelling of the system, especially in complex systems; it may be challenging to exactly model the system under study [21]. Even it does not require the partial model of the system as some intelligent optimal control methods such as reinforcement learning need [22,23]. ESC methods attracted significant interest among control theory researchers after the first stability and convergence proof was performed. The ESC methods can be classified from different perspectives such as stochastic and deterministic for the search signal. The performance of ESC methods is very sensitive concerning adjustable parameters such as feedback gain, and practical tuning of the coefficient is complicated [24]. Based on [25], ESC methods can be classified into five main categories, including sliding mode-based ESC, neural network-based ESC, adaptive-based ESC, AESC, and finally, perturbation-based ESC in which each of the mentioned methods can be implemented in both unidirectional and multidimensional problems. After a brief explanation of each method, Ref. [25] has investigated them for robotic applications.

Recently, ESC methods have been implemented in several practical and theoretical applications, including performing maximum output flow for rotatory blood pumps in a left ventricular assist device [26], maximizing output power of an independent autonomous excavator [27], stiffness auto-tuning of quasi-passive ankle exoskeleton for real-time minimization of muscles effort [28], fast and robust real-time energy management of fuel cell hybrid electric vehicles utilizing fractional order ESC [29], formation control of 1,200 high-altitude balloons which provide Internet for ground users [30], maximizing friction of an active braking system implemented on electric vehicles [31] and minimization of power consumption in vapor compression systems implemented on commercial vehicles for vehicle owner comfort [32]. In combination with AESC, a high-gain step-up converter was utilized in this chapter to track the MPP. Nowadays, series and parallel combinations of solar panels are utilized in practice to provide high voltages for load demand. However, this leads to reduced efficiency, higher implementation costs, and increased size of the implemented system. Thus, high-gain converters provide high voltage rates for microgrids and electric vehicles [33]. The utilized high gain converter in this chapter has several advantages, including high-voltage conversion gain, continuous input current, simple structure, high ratio of step-up gain in comparison with used components, common ground between source and load, simple control, and minimum operating modes, less cost, size and weight and ability of MPPT in PV systems. Other sections are structured as follows. First, the suggested structure is represented, then, the mathematical formulation of solar cells, high gain DC–DC converter are investigated in Section 3.2. The proposed MPPT method is investigated in Section 3.3. Next, in Section 3.4, the suggested structure is investigated through simulations based on the mathematical model, and finally, the chapter is concluded in the conclusion section.

3.2 System description

Figure 3.1 depicts the under-study system, which consists of PV array, DC–DC converter, load, and AESC controller. This system has been considered for performance verification of the proposed (AESC) on the MPPT of the PV panel. The system has been precisely modeled by taking the impacts of both irradiation and ambient temperature into consideration. The step-by-step explanation of the considered system comes in the following.

Step 1: The output voltage (V_{pv}) and current (I_{pv}) of PV array are measured by sensors.

Step 2: The output power of PV array (P_{pv}) is calculated through $P_{pv} = V_{pv}.I_{pv}$.

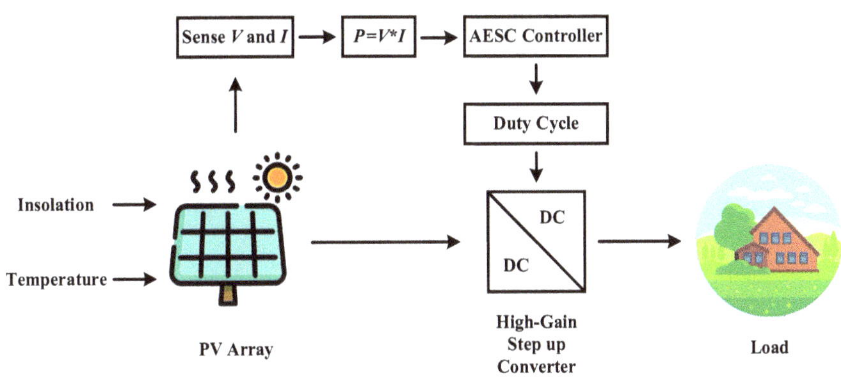

Figure 3.1 The schematic of the considered system employing the proposed AESC controller

Step 3: The calculated power is sent to the AESC block and the gradient information of the system is collected.

Step 4: The control signal provided by the AESC block is applied to the DC–DC converter through the switches' duty cycle. The AESC controller regulates the duty cycle of switches such that the maximum power is extracted from the PV array. It is noted that the control constraints maintain the system at the extremum point.

3.2.1 Photovoltaic array

Solar energy emitted from the sun can be converted to electrical energy through PV arrays. The semiconductor-based p–n junction is utilized to absorb solar energy. Series and parallel combinations of single cells are used to form PV modules to provide higher voltage and current ratings for supplying load demands. Several modules can be combined to provide higher output powers. PV arrays have a nonlinear nature. For easy implementation, the utilized model for PV arrays must be simple, the direct effect of ambient temperature and insolation must be considered, and the model must be connected to real-world parameters through the given data sheets with the PV array. The equivalent circuit of a single PV cell is represented in Figure 3.2 using a current source paralleled with an ideal diode. These two are serried with a resistor that describes the internal resistance of wires and junctions [34].

The *I–V* characteristic of solar cells can be represented by (3.1):

$$I = I_p - I_s\left(\exp\frac{V + IR_s}{\varepsilon V_t} - 1\right) \tag{3.1}$$

where the I_p, I_s, V, ε, V_t, and T represent the photocurrent of each cell, reverse saturation current of the cell, applied voltage on diode terminals, solar cell ideality factor, the thermal voltage that is represented by $V_t = \frac{k_B T}{q}$ where k_B is the Boltzmann constant, T is diode absolute temperature in the Kelvin unit and q is the electron charge, respectively. Generated photocurrent of each cell linearly depends on the insolation level, and the ambient temperature also affects it, as well. Eqs. (3.2) and (3.3) represent the related equations:

$$I_p = I_{pr} + K_0(T - T_r) \tag{3.2}$$

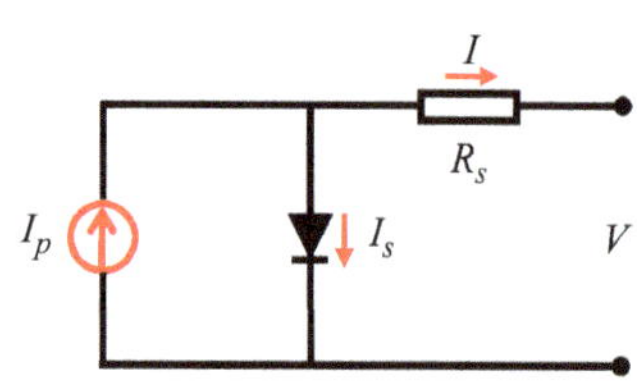

Figure 3.2 The equivalent circuit of a single PV cell [34]

$$I_{pr} = I_{sc_r}\frac{E}{E_r} \tag{3.3}$$

where r, E, and K_0 indicate reference value, insolation, and short circuit temperature coefficients, respectively. I_{sc_r}, T_r, and E_r can be attained from the PV data sheet performed by the manufacturer. The reverse saturation current of the solar cell can be represented by (3.4) and (3.5):

$$I_s = I_{s_r}\left(\frac{T}{T_r}\right)^{\frac{3}{\varepsilon}}\exp\left(-\frac{qV_g}{\varepsilon k_B}\left(\frac{1}{T}-\frac{1}{T_r}\right)\right) \tag{3.4}$$

$$I_{s_r} = \frac{I_{sc_r}}{exp\left(\frac{qV_{ocr}}{\varepsilon k_B T_r}\right)-1} \tag{3.5}$$

where V_g represents the semiconductor band gap voltage. Finally, the solar cell series resistance can be represented by (3.6) and (3.7):

$$R_s = -\frac{dV}{dI_{V_OC}} - \frac{1}{X_V} \tag{3.6}$$

$$X_V = I_{s_r}\frac{q}{\varepsilon k_B T_r}\exp\left(\frac{qV_{OC_r}}{\varepsilon k_B T_r}\right) \tag{3.7}$$

where the parameter $\frac{dV}{dI_{V_OC}}$ can be calculated from performed data by the manufacturer [34].

3.2.2 Suggested high-gain DC–DC converter

One of the most critical devices of the suggested structure is the DC–DC converter which acts as a power interface between the PV array and the load. A utilized DC–DC converter must have several advantages like high gain, high conversion efficiency, ease of control, and simple operational structure. Due to the worldwide popularity of microgrids and distributed energy resources, DC–DC converters gained lots of interest, especially high-gain DC–DC converters with novel structures [35]. This chapter utilized a high gain non-isolated non-coupled inductor-based structure to provide MPPT. The proposed structure in [35] by Varesi *et al.*, as seen in Figure 3.3, consists of two switches, three capacitors, three diodes, and two inductors. The low number of utilized components leads to fewer power losses due to decreasing the parasitic elements of switches, diodes, capacitors, and inductors. The utilized switches turn on/off simultaneously, which leads to easy control of the converter. The input current of converter is divided between inductors, leading to high conversion efficiency and lower power losses.

The converter's continuous input current allows it to perform MPPT in PV applications. Therefore, continuous conduction mode (CCM) analysis is essential for MPPT applications. The utilized structure has two operating modes in continuous conduction mode. In mode 1, as can be seen in Figure 3.4, two switches S_1 and S_2 are turned on, D_O and D_1 are reversed biased while the D_2 is forward-biased.

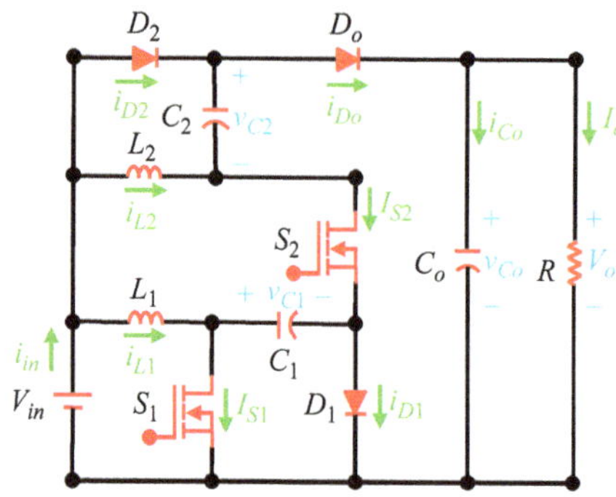

Figure 3.3 General structure of utilized high-gain DC–DC converter [35]

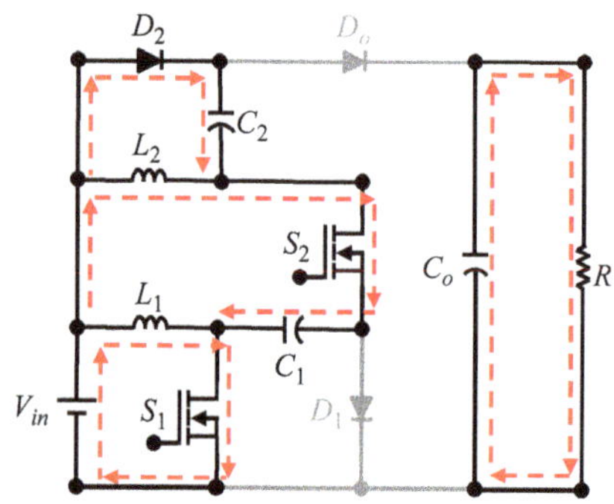

Figure 3.4 High-gain DC–DC converter in continuous conduction mode (mode 1) [35]

Input DC voltage source V_{in} charges the L_1 while C_1 is discharged. L_2 and C_2 are charged by V_{in} and C_1 and C_O provides energy for the output load. The governing equations of the first mode can be represented as follows:

Assuming the capacitor voltages ($v_{C_O}, v_{C_1}, v_{C_2}$) and inductors' current (i_{L_1}, i_{L_2}) as the state-space variables, the state-space representation of mode 1 can be achieved as:

$$\begin{cases} v_{L_1} = L_1 \dfrac{di_{L_1}}{dt} = v_{in} \\ v_{L_2} = L_2 \dfrac{di_{L_2}}{dt} = v_{C_2} \\ i_{C_1} = C_1 \dfrac{dv_{C_1}}{dt} = i_{L_1} - i_{in} \\ i_{C_2} = C_2 \dfrac{dv_{C_2}}{dt} \\ i_{C_O} = C_O \dfrac{dv_{C_O}}{dt} = -\dfrac{v_O}{R} \\ i_{in} = i_{L_1} + i_{L_2} + \left(\dfrac{v_{C_2}}{r_{D_2} + r_{C_2}}\right) \end{cases} \tag{3.8}$$

$$\begin{bmatrix} L_1 & 0 & 0 & 0 & 0 \\ 0 & L_2 & 0 & 0 & 0 \\ 0 & 0 & C_1 & 0 & 0 \\ 0 & 0 & 0 & C_2 & 0 \\ 0 & 0 & 0 & 0 & C_O \end{bmatrix} \frac{d}{dt} \begin{bmatrix} i_{L_1}(t) \\ i_{L_2}(t) \\ v_{C_1}(t) \\ v_{C_2}(t) \\ v_{C_O}(t) \end{bmatrix} = \underbrace{\begin{bmatrix} 0 & 0 & 0 & 0 & 0 \\ 0 & 0 & 0 & 1 & 0 \\ 1 & 0 & 0 & 0 & 0 \\ -1 & -1 & 0 & 0 & 0 \\ 0 & 0 & 0 & 0 & -\frac{1}{R} \end{bmatrix}}_{A_1} \begin{bmatrix} i_{L_1} \\ i_{L_2} \\ v_{C_1} \\ v_{C_2} \\ v_{C_O} \end{bmatrix} + \underbrace{\begin{bmatrix} 1 \\ 0 \\ -1 \\ 1 \\ 0 \end{bmatrix}}_{B_1} v_{in} \tag{3.9}$$

$$y(t) = \underbrace{\begin{bmatrix} 1 & 1 & 0 & \dfrac{1}{r_{D_1} + r_{C_2}} & 0 \end{bmatrix}}_{E_2} \begin{bmatrix} i_{L_1}(t) \\ i_{L_2}(t) \\ v_{C_1}(t) \\ v_{C_2}(t) \\ v_{C_O}(t) \end{bmatrix} + \underbrace{[0]}_{F_1} V_{in} \tag{3.10}$$

In the second mode, as can be seen in Figure 3.5, the switches are turned off, D_2 is reversed-biased, D_O and D_1 are forward-biased. In this mode, V_{in} and L_1 charges C_1 and V_{in}, L_2, C_2 supply the load and C_O. The following equations represent the operating conditions of the converter in the second mode.

Similar to mode 1, the state-space representations of the second mode can be represented as follows:

$$\begin{cases} v_{L_1} = L_1 \dfrac{di_{L_1}}{dt} = v_{in} - v_{C_1} \\ v_{L_2} = L_2 \dfrac{di_{L_2}}{dt} = v_{in} + v_{C_2} - v_O \\ i_{C_1} = C_1 \dfrac{dv_{C_1}}{dt} = i_{L_1} \\ i_{C_2} = C_2 \dfrac{dv_{C_2}}{dt} = -i_{L_2} \\ i_{C_O} = C_O \dfrac{dv_{C_O}}{dt} = i_{L_2} - \dfrac{v_O}{R} \\ i_{in} = i_{L_1} + i_{L_2} \end{cases} \tag{3.11}$$

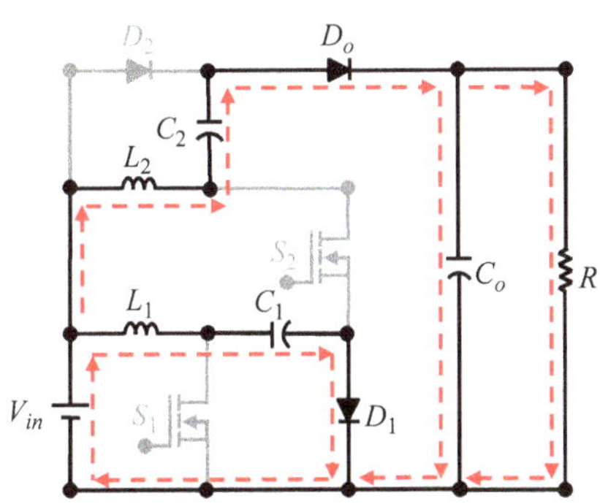

Figure 3.5 High-gain DC–DC converter mode 2 in continuous conduction mode [35]

$$\begin{bmatrix} L_1 & 0 & 0 & 0 & 0 \\ 0 & L_2 & 0 & 0 & 0 \\ 0 & 0 & C_1 & 0 & 0 \\ 0 & 0 & 0 & C_2 & 0 \\ 0 & 0 & 0 & 0 & C_O \end{bmatrix} \frac{d}{dt} \begin{bmatrix} i_{L_1}(t) \\ i_{L_2}(t) \\ v_{C_1}(t) \\ v_{C_2}(t) \\ v_{C_O}(t) \end{bmatrix} = \underbrace{\begin{bmatrix} 0 & 0 & -1 & 0 & 0 \\ 0 & 0 & 0 & 1 & -1 \\ 1 & 0 & 0 & 0 & 0 \\ 0 & -1 & 0 & 0 & 0 \\ 0 & 1 & 0 & 0 & -\frac{1}{R} \end{bmatrix}}_{A_2} \begin{bmatrix} i_{L_1} \\ i_{L_2} \\ v_{C_1} \\ v_{C_2} \\ v_{C_O} \end{bmatrix} + \underbrace{\begin{bmatrix} 1 \\ 1 \\ 0 \\ 0 \\ 0 \end{bmatrix}}_{B_2} [v_{in}] \tag{3.12}$$

$$y(t) = [i_{in}(t)] = \underbrace{[1 \quad 1 \quad 0 \quad 0 \quad 0]}_{E_2} \begin{bmatrix} i_{L_1} \\ i_{L_2} \\ v_{C_1} \\ v_{C_2} \\ v_{C_O} \end{bmatrix} + \underbrace{[0]}_{F_2} [v_{in}] \tag{3.13}$$

By the combination of state-space representation of modes 1 and 2, assuming $\dot{x} = \bar{A}x + \bar{B}u$ and $y = \bar{C}x + \bar{D}u$, the general state-space representation of the high-gain DC–DC converter can be represented as follows:

$$A = DA_1 + D^{'}A_2 = D\begin{bmatrix} 0 & 0 & 0 & 0 & 0 \\ 0 & 0 & 0 & 1 & 0 \\ 1 & 0 & 0 & 0 & 0 \\ -1 & -1 & 0 & 0 & 0 \\ 0 & 0 & 0 & 0 & -\frac{1}{R} \end{bmatrix} + D^{'}\begin{bmatrix} 0 & 0 & -1 & 0 & 0 \\ 0 & 0 & 0 & 1 & -1 \\ 1 & 0 & 0 & 0 & 0 \\ 0 & -1 & 0 & 0 & 0 \\ 0 & 1 & 0 & 0 & -\frac{1}{R} \end{bmatrix}$$

$$= \begin{bmatrix} 0 & 0 & -D^{'} & 0 & 0 \\ 0 & 0 & 0 & 1 & -D^{'} \\ 1 & 0 & 0 & 0 & 0 \\ -D & -1 & 0 & 0 & 0 \\ 0 & D^{'} & 0 & 0 & -\frac{1}{R} \end{bmatrix} \tag{3.14}$$

$$B = DB_1 + D^{'}B_2 = D\begin{bmatrix} 1 \\ 0 \\ -1 \\ 1 \\ 0 \end{bmatrix} + D^{'}\begin{bmatrix} 1 \\ 1 \\ 0 \\ 0 \\ 0 \end{bmatrix} = \begin{bmatrix} 1 \\ D^{'} \\ -D \\ D \\ 0 \end{bmatrix} \tag{3.15}$$

$$E = DE_1 + D^{'}E_2 = D\left[1 \quad 1 \quad 0 \quad \frac{1}{r_{D_2} + r_{C_2}} \quad 0\right] + D^{'}[1 \quad 1 \quad 0 \quad 0 \quad 0]$$

$$= \left[1 \quad 1 \quad 0 \quad \frac{D}{r_{D_2} + r_{C_2}} \quad 0\right] \tag{3.16}$$

$$F = DF_1 + D^{'}F_2 = D[0] + D^{'}[0] = 0 \tag{3.17}$$

$$\begin{bmatrix} L_1 & 0 & 0 & 0 & 0 \\ 0 & L_2 & 0 & 0 & 0 \\ 0 & 0 & C_1 & 0 & 0 \\ 0 & 0 & 0 & C_2 & 0 \\ 0 & 0 & 0 & 0 & C_O \end{bmatrix} \frac{d}{dt} \begin{bmatrix} i_{L_1}(t) \\ i_{L_2}(t) \\ v_{C_1}(t) \\ v_{C_2}(t) \\ v_{C_O}(t) \end{bmatrix} = \begin{bmatrix} 0 & 0 & -D' & 0 & 0 \\ 0 & 0 & 0 & 1 & -D' \\ 1 & 0 & 0 & 0 & 0 \\ -D & -1 & 0 & 0 & 0 \\ 0 & D' & 0 & 0 & -\frac{1}{R} \end{bmatrix} \begin{bmatrix} i_{L_1} \\ i_{L_2} \\ v_{C_1} \\ v_{C_2} \\ v_{C_O} \end{bmatrix} + \begin{bmatrix} 1 \\ D' \\ -D \\ D \\ 0 \end{bmatrix} [v_{in}] \tag{3.18}$$

$$[i_{in}(t)] = \begin{bmatrix} 1 & 1 & 0 & \frac{D}{r_{D_2} + r_{C_2}} & 0 \end{bmatrix} \begin{bmatrix} i_{L_1} \\ i_{L_2} \\ v_{C_1} \\ v_{C_2} \\ v_{C_O} \end{bmatrix} \tag{3.19}$$

In CCM, based on the volt-second balance principle, the converter gain can be represented by (3.20), where the parameter D represents the duty cycle of the switches [35]:

$$\langle v_{L_i} \rangle = 0 \Rightarrow \begin{cases} v_{C_1} = \dfrac{V_{in}}{(1-D)} \\ v_{C_2} = \dfrac{(2-D)V_{in}}{(1-D)} \end{cases} \Rightarrow G_{CCM} = \frac{V_O}{V_{in}} = \frac{D^2 - 3D + 3}{(1-D)^2} \tag{3.20}$$

3.3 Proposed AESC technique

In this section, the AESC approach for MPPT of solar array utilizing suggested high gain DC–DC converter with control design considerations is investigated in detail through governing mathematical equations. Assume a general linear time-invariant system (SISO) represented as (3.21) and a performance function as (3.22):

$$\dot{x} = Ax + Bu \tag{3.21}$$

$$z = F(x) \tag{3.22}$$

in which $x \in \mathbb{R}^n$, $u \in \mathbb{R}$, $z \in \mathbb{R}$, and $F : \mathbb{R}^n \to \mathbb{R}$ are the states of the system, the system input, the performance output of the considered system, and a continuously differentiable function, respectively. A and B are the matrixes of the system, nevertheless, the explicit form of F and the maximum point is unknown. To explain this framework more clearly, the discussion is separated into two parts: first, searching for the maximum point, and second, designing a suitable control effort to maintain and track the maximum point [36]. It is clear that the MPPT problem is a maximization one, but since usually optimization problems work with minimization, by defining the performance function as $\bar{z} := -F(x) := \mathcal{J}(x)$ the problem is transformed into the standard form of minimization problems. The Block diagram of the AESC method is represented in Figure 3.6. In this figure, the block entitled "Numerical Optimization Algorithm" will be designed to solve the first part of the problem, and the "state

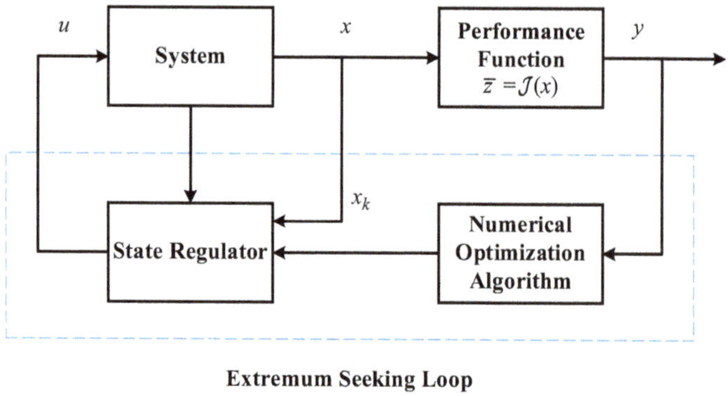

Figure 3.6 Block diagram of AESC [37]

regulator" block is considered to force the system to follow the maximum point utilizing the control effort u.

Two assumptions are made about the assumed system (3.21).

Assumption 3.1 *The performance function $\mathcal{J}(x)$ is generally unavailable, but it is continuously differentiable, bounded, and in a convex form.*

Assumption 3.2 *The considered system is stable and controllable.*

The ESC problem is defined as a numerical optimization as (3.23):

$$\min_{x \in \mathbb{R}^n} \quad \mathcal{J}(x) \quad \text{subject to } \dot{x} = Ax + Bu \tag{3.23}$$

which is a constrained minimization problem and is different from the custom algebraic constraints. The systems state vector x is feasible if it represents a solution for the considered dynamic system. With a controllable pair of (A, B), the control input u transfer state x to any other point in $\mathbb{R}^n$ in a finite time always exists [38], which demonstrates the requirement for the second assumption. Controllable constraints of the considered dynamic system allow state x to be anywhere in the state space where the numerical optimizer sets and the procedure that state x reaches the specific point is defined by the dynamic system [36].

3.3.1 Line search-based optimization methods

Investigated approach to seek the extremum for MPPT application is based on the line search method proposed in [36,39,40]. Therefore, this section has been represented based on mentioned references. In the minimization problem, $\min_{x \in \mathbb{R}^n} \mathcal{J}(x)$ which is unconstrained, a suitable search direction $d_k \in \mathbb{R}^n$ is calculated in each iteration of the line search method, next, the method decides the move length in the calculated direction. Mentioned iteration of the line search method is represented by (3.24):

$$x_{k+1} = x_k + \mathfrak{s}_k d_k \tag{3.24}$$

in which $\mathfrak{s}_k$ represents the step length change and needs d_k to be a descent direction, one direction which $d_k^{\mathrm{T}} \nabla \mathcal{J}(x_k) < 0$, because mentioned feature guarantees the performance function $\mathcal{J}$ reduction along the calculated direction. The most apparent option for search direction is the steepest descent direction which can be represented by $d_k = -\nabla \mathcal{J}(x_k)$. Applying a decent change direction d_k, a tradeoff is faced in selecting the step change length $\mathfrak{s}_k$ which leads to a remarkable reduction of performance function $\mathcal{J}$ and the selection of step length is not time-consuming. By approximately solving the following one-dimensional optimization problem, the proper step length can be calculated by (3.25):

$$\min_{\mathfrak{s}>0} \phi(\mathfrak{s}) = \min_{\mathfrak{s}>0} \mathcal{J}(x_k + \mathfrak{s} d_k) \tag{3.25}$$

It is costly and inessential to solve an exact optimization of $\phi(\mathfrak{s})$ to find $\mathfrak{s}$. An inexact line search method to recognize a step length that reaches a satisfactory reduction in performance function $\mathcal{J}$ at optimal cost is carried out by more practical approaches. In specific, the Armijo condition is represented as follows:

$$\mathcal{J}(x_k + \mathfrak{s}_k d_k) \leq \mathcal{J}(x_k) + \eta_1 \mathfrak{s}_k d_k^{\mathrm{T}} \nabla \mathcal{J}(x_k) \tag{3.26}$$

prevents long steps using a significant decrease criterion, on the other hand, the Wolfe condition, (3.27)

$$d_k^{\mathrm{T}} \nabla \mathcal{J}(x_k + \mathfrak{s}_k d_k) \geq \eta_2 d_k^{\mathrm{T}} \nabla \mathcal{J}(x_k) \tag{3.27}$$

prevents short steps utilizing curvature criterion, with $0 < \eta_1 < \eta_2 < 1$. The constraint $\eta_2 > \eta_1$ guarantees the existence of acceptable points. Moreover, to avoid a weak selection of descent directions, an angle condition is considered to force a uniform lower bound on the angle between d_k and $-\nabla \mathcal{J}(x_k)$, which can be represented as follows [36]:

$$\cos \alpha_k = \frac{-d_k^{\mathrm{T}} \nabla \mathcal{J}(x_k)}{\|d_k\| \|\nabla \mathcal{J}(x_k)\|} \geq \eta_3 > 0 \tag{3.28}$$

in which η_3 does not depend on k. The following represents the first-order global convergence results for line search-based methods [41].

Theorem 3.1: [42,43] *Assume that the considered convex performance function $\mathcal{J} : \mathbb{R}^n \to \mathbb{R}$ be bounded below and continuously differentiable on $\mathbb{R}^n$. Consider that $\nabla \mathcal{J}$ is continuously Lipschitz with defined constant $\mathcal{L}$, that is, $\|\nabla \mathcal{J}(y) - \nabla \mathcal{J}(x)\| \leq \mathcal{L} \|y - x\|$ for all $x, y \in \mathbb{R}^n$. If the sequence $\{x_k\}$ meets the considered conditions (3.26)–(3.28), then*

$$\lim_{k \to \infty} \|\nabla \mathcal{J}(x_k)\| = 0 \tag{3.29}$$

In [36,42], it is proved that for any $x, y \in \mathbb{R}^n$ and Lipschitz continuous function $\nabla \mathcal{J}$ with constant $\mathcal{L}$, we have the following inequalities:

$$\mathcal{J}(x + y) \leq \mathcal{J}(x) + y^{\mathrm{T}} \nabla \mathcal{J}(x) + \frac{\mathcal{L}}{2} \|y\|^2 \tag{3.30}$$

$$\mathcal{J}(x_k + \mathfrak{s}_k d_k) - \mathcal{J}(x_k) \leq -\frac{c}{2\mathcal{L}} \|\nabla \mathcal{J}(x_k)\|^2 \cos^2 \alpha_k \tag{3.31}$$

where $\mathfrak{s}_k, d_k$ are the step length and descent direction respectively, $\eta = 1$ for exact line search, and $\eta = 2\eta_1(1 - \eta_2)$ for exact line search fulfilling conditions (3.26) and α_k is the angle between the vector d_k and $-\nabla \mathcal{J}(x_k)$. Hence for ensuring the feasibility of an inexact line search η_1 and η_2 must satisfy $0 < \eta_1 < \eta_2 < 1$, $\eta = 2\eta_1(1 - \eta_2) < 1$ is obtained. This remark is consistent for upper bound results in two mentioned inequalities (3.30) and (3.31). Hence, it is always expected that an exact line search method to have more reduction along the search direction than an inexact one [36].

3.3.2 Control scheme

In solar-cell control approaches, it is necessary to track the MPP that varies based on the ambient conditions. The main objective in the design of the controller of these systems is to perform a control effort u that derives the system states from the previous MPP (initial state x_0) to the new MPP x_1 in a finite time. The following theorem of [38] provides sufficient conditions for the existence of such a control effort u for LTI systems.

Theorem 3.2: [38] *Assume system (3.21) with Assumption 3.2, then for any $x(t_0) = x_0$ and any x_1 there exists a control input u that derives x_0 to x_1 in a finite time. Furthermore, for the single-input LTI systems, this control input can be represented as follows:*

$$u(t) = -B^{\mathrm{T}} e^{A^{\mathrm{T}}(t_1 - t)} \mathcal{W}_c^{-1}(t_1)[e^{A(t_1 - t_0)} x_0 - x_1] \tag{3.32}$$

which will derive the initial state x_0 to final state x_1 at time t_1, in which $\mathcal{W}_c$ represents the controllability Gramian and can be represented as (3.33):

$$\mathcal{W}_c(t_1) = \int_{t_0}^{t_1} e^{A(t_1 - \tau)} BB^{\top} e^{A^{\top}(t_1 - \tau)} d\tau = \int_0^{t_1 - t_0} e^{A\tau} BB^{\top} e^{A^{\top}\tau} d\tau \tag{3.33}$$

In the proposed AESC approach, the state tracking control is utilized (3.32) for MPPT of the solar array.

3.3.3 Extremum seeking control approach

For the system (3.21) as a controllable LTI system, a line search method is studied and the controller is designed for state tracking of the considered LTI system (3.21) for developing an ESC-based framework that independently optimizes $y = \mathcal{J}(x)$. The principle of this framework can be expressed as the following pseudo-code (3.1).

3.3.4 Convergence analysis of the AESC approach

Theorem 3.3: *Consider that $\dot{x} = Ax + Bu$ is controllable and $\mathcal{J} : \mathbb{R}^n \to \mathbb{R}$ is continuously recognizable on $\mathbb{R}^n$ bounded below and is in the convex form. Moreover, consider that $\nabla \mathcal{J}$ is continuously Lipschitz with defined constant $\mathcal{L}$. If the represented ES framework is applied, the convex performance function will converge to the global optimal point as $t \to \infty$* [36].

Proof. Based on the ES approach, line search methods with first global convergence feature will perform a descent sequence $\{x_k\}$ which converges to the global optimal point of the represented convex performance function as $k \to \infty$. Proposed controller (3.34) interpolates between the $\{x_k\}$ precisely within a finite time δ_k from state x_k to state x_{k+1}, so the convex performance function reaches the global optimal point as $t \to \infty$.

Step *:
Given $x_0, t_0 = 0$
Set ε_0 & $k := 0$

Step **:
To produce $x_{k+1} = x_k + \mathfrak{s}_k d_k$, utilize an exact/inexact line search approach, in which, $\{x_k\}$ will converge to the optimal point of the performance function, globally. If $\|\nabla \mathcal{J}(x_k)\| < \varepsilon_0$, then stop.

Step *:**
Select δ_k, consider $t_{k+1} = t_k + \delta_k$, in time step $t_k \le t \le t_{k+1}$ the applied input is

$$u(t) = -B^\top e^{A^\top (t_{k+1}-t)} \mathcal{W}_c^{-1}(t_{k+1})[e^{A\delta_k} x_k - x_{k+1}] \tag{3.34}$$

in which $\mathcal{W}_c(t_{k+1})$ can be represented as

$$\mathcal{W}_c(t_{k+1}) = \int_0^{\delta_k} e^{A\tau} BB^\top e^{A^\top \tau} d\tau$$

Step **:**
Define $k \leftarrow k + 1$.
Goto **Step** **.

Algorithm 3.1: Overall representation of the ESC approach [36]

Corollary 3.1: *Extra to the considerations in Theorem 3.3, let x^* represent the not-known global optimal point of the convex performance function. If $\mathcal{J}$ is strongly convex on $\mathbb{R}^n$, and the steepest descent approach with exact line search is utilized in the ES approach, subsequently, the system states will converge to the ε vicinity of x^* most time $t = \sum_{k=1}^{N} \delta_k$, that $N = \frac{\log_{10}(\frac{f(x_0)-f(x^*)}{\varepsilon})}{\log_{10}(\frac{1}{h})}$ for some $0 < h < 1$ [36].*

Proof. If the convex performance function $\mathcal{J}$ is strongly convex on $\mathbb{R}^n$, which denotes there exist defined constants $Q, q > 0$ such that [44]

$$qI \le \nabla^2 f(x) \le QI \tag{3.35}$$

for all $x \in \mathbb{R}^n$. If the steepest descent framework is utilized in the ES approach, letting $h = 1 - \frac{q}{Q} < 1$, it must lead to $f(x_k) - f(x^*) \le \varepsilon$ after

$$N = \frac{\log_{10}(\frac{(f(x_0)-f(x^*))}{\varepsilon})}{\log_{10}(\frac{1}{h})} \tag{3.36}$$

iterations of the steepest descent framework with exact line search, subsequently, the system states x will converge to the ε vicinity of x^* at most $t = \sum_{k=1}^{N} \delta_k$ [36].

For better evaluation of the proposed approach, simulation results have been conducted for the conventional P&O method, too. The conventional P&O method utilizes fixed-step changes in the duty cycle for MPPT of PV systems and can be implemented in various applications. It is the most popular MPPT approach despite its weaknesses and in lots of papers, it is implemented for comparison of various methods. It has a simple structure and can be implemented in real-time applications and commercial ones. P&O's working principle can be represented as pseudo-code 3.2 [45].

3.4 Simulation and comparison results

The performance of the proposed AESC method is evaluated by simulation results conducted in MATLAB–Simulink. A 730 W solar module has been employed. Also, it has been assumed that all the components are ideal, having linear time-invariant responses. For better evaluation, the conventional P&O method has also been investigated through simulations and has been compared with the proposed method. All the simulation results have been obtained by sampling period of 0.01ms. Two scenarios have been considered. In the first scenario, the ambient temperature is kept constant, but sudden changes occur in solar irradiation. In the second scenario, the solar irradiation is kept constant and step-changes happen at ambient temperature. The results for each scenario come in the following.

3.4.1 Scenario 1

Different irradiation values have been applied to the solar module and the output powers tracked by proposed AESC and conventional P&O methods have been investigated. The ambient temperature has been considered to be 25°C. The irradiation input of the solar module varies with time-based on step changes in 0.5 s periods, and the total simulation time of the system is set to be 3s, containing six steps. Therefore, irradiation step changes with values of 800–700–1,000–900–600–800, as represented in Figure 3.7 have been applied to the solar module. In addition, various irradiation values are applied to the detailed investigation of the proposed method in different weather conditions. Figure 3.8 shows the current variations caused by the irradiation inputs. The curve has the minimum ripples in the steady-state response with the proposed AESC approach. The steady-state ripples are essential in solar PV systems because if the ripples are not dealt with in the steady-state, the lifespan of both the solar array and the DC–DC converter will be reduced. Hence, it is clear that the proposed control strategy has a remarkable performance in reducing steady-state oscillations. Figure 3.9 depicts the power–voltage curve when irradiation inputs are applied to the PV system. This diagram is fundamental in designing the MPPT controller for the PV system. According to this figure, the MPP is obtained where the derivative of the output power to the voltage equals zero (i.e. $\frac{\partial P_{pv}}{\partial V_{pv}} = 0$). Hence, the operating point keeps moving till it reaches

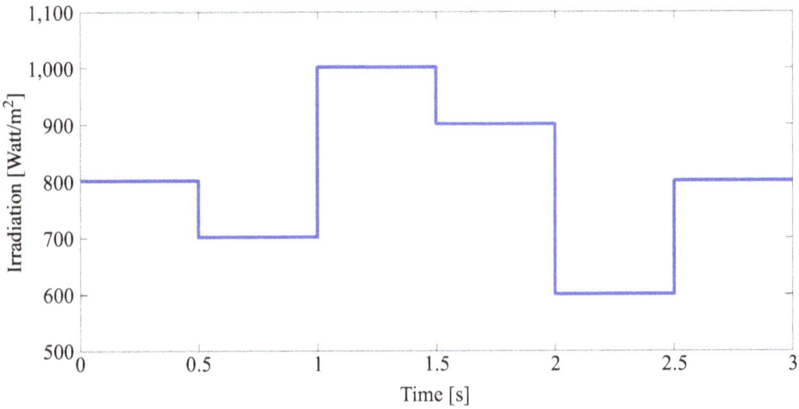

Figure 3.7 Irradiation inputs applied to the PV system in scenario 1

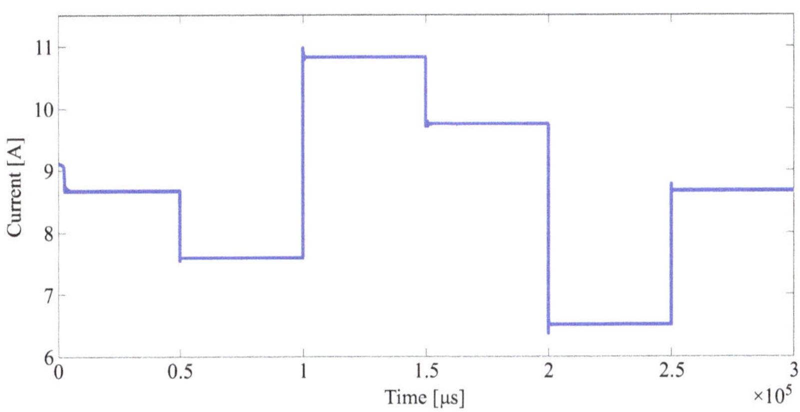

Figure 3.8 PV Current variations in response to the irradiation changes

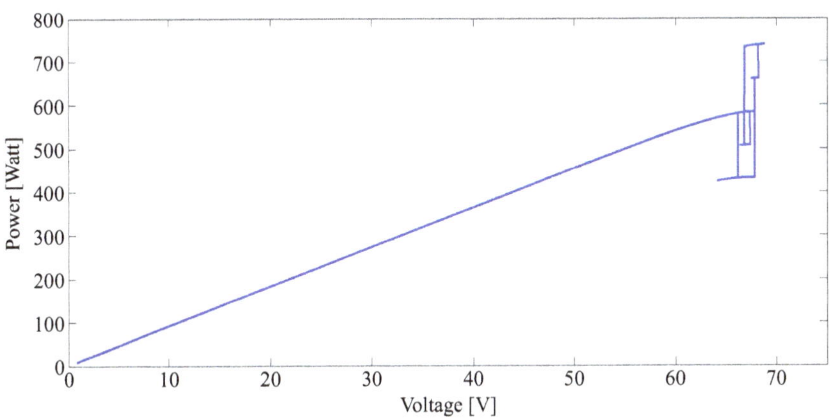

Figure 3.9 Power-voltage characteristic of the PV system in scenario 1

the MPP. Then, the controller keeps the system operating around the MPP. The current–voltage curve is represented in Figure 3.10 which shows that the AESC can track the MPP of the PV system and then regulates the system around this point. Therefore, the current–voltage curve is a fundamental characteristic of the power–voltage curve of the PV system. Figure 3.11 demonstrates the output power tracked by the proposed AESC and the conventional P&O methods. As seen from the zoomed figure, the output power resulting from the proposed AESC method has smaller oscillations in the MPP than that of the conventional P&O method. Also, the suggested method provides a faster response than the conventional P&O method and reaches the MPP in a shorter time than the conventional P&O method. In addition, the simulation run-time of the suggested method is much shorter than that of the conventional P&O method.

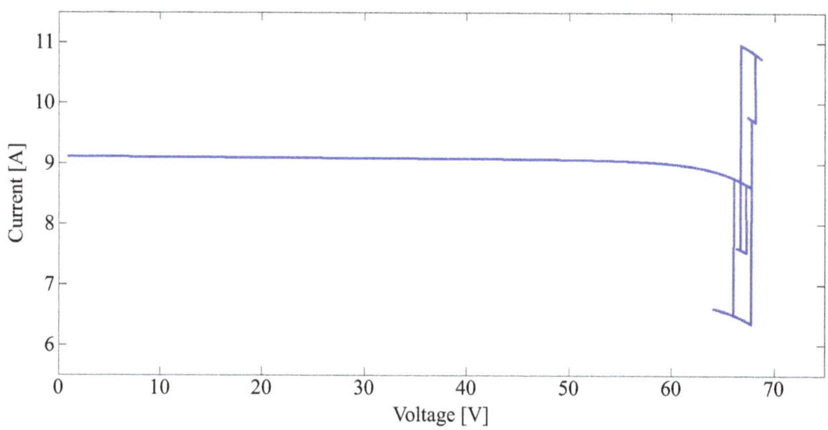

Figure 3.10 Current–voltage curve in scenario 1

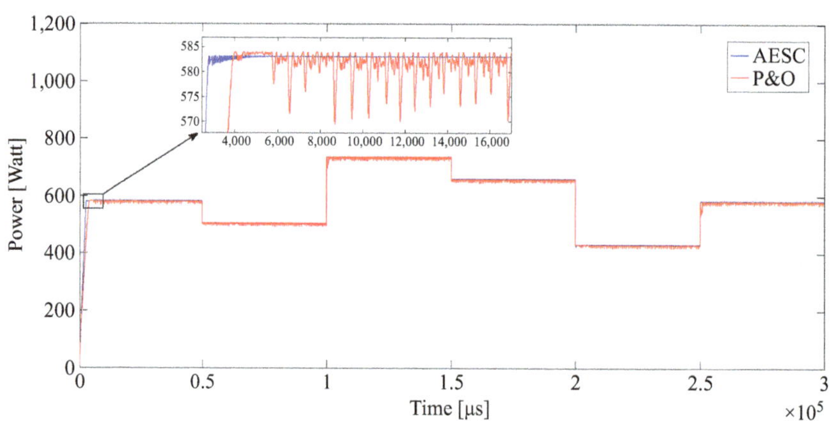

Figure 3.11 Tracked output power of proposed AESC and conventional P&O methods during sudden changes in solar irradiation

Input data: V_{PV}, I_{PV}

Output data: V_{ref}

```
*
Sense V_PV & I_PV
P_PV (i) ← V_PV (i) × I_PV (i)
if P_PV(i) − P_PV (i−1) = 0 then
|   goto *
else
|   goto **
end
**
if P_PV (i) − P_PV (i−1) > 0 then
|   if V_PV (i) − V_PV (i−1) > 0 then
|   |   V_PV (i) = V_PV (i) − Δ
|   |   goto *
|   else
|   |   V_PV (i) = V_PV (i) + Δ
|   |   goto *
|   end
else
|   goto ***
end
***
if V_PV (i) − V_PV (i−1) > 0 then
|   V_PV (i) = V_PV (i) − Δ
|   goto *
else
|   V_PV (i) = V_PV (i) + Δ
|   goto *
end
```

Algorithm 3.2: P&O algorithm for the PV system [45]

3.4.2 Scenario 2

In the second scenario, various ambient temperature values have been considered, while the solar irradiation has been kept constant on 750 W/m^2. The output powers resulting from the proposed AESC and the conventional P&O methods have been studied. Ambient temperature step-changes with values of 29–31–35–22–15–18 with a period of 0.5 s with a total simulation time of 3 s which contains six steps have been applied to the PV system, which can be represented as Figure 3.12. Similar to scenario 1, current variations caused by the temperature inputs are represented in Figure 3.13. The current curve has the minimum ripples in the steady state. The proposed AESC approach remarkably reduced the steady-state

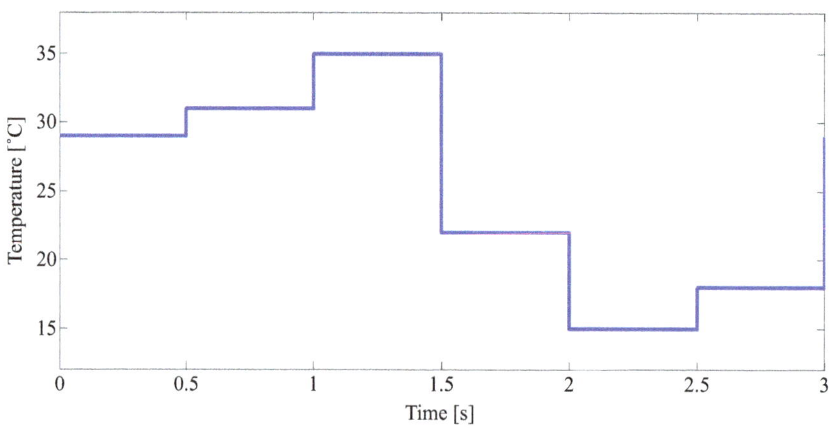

Figure 3.12 Temperature inputs given to the PV system in scenario 2

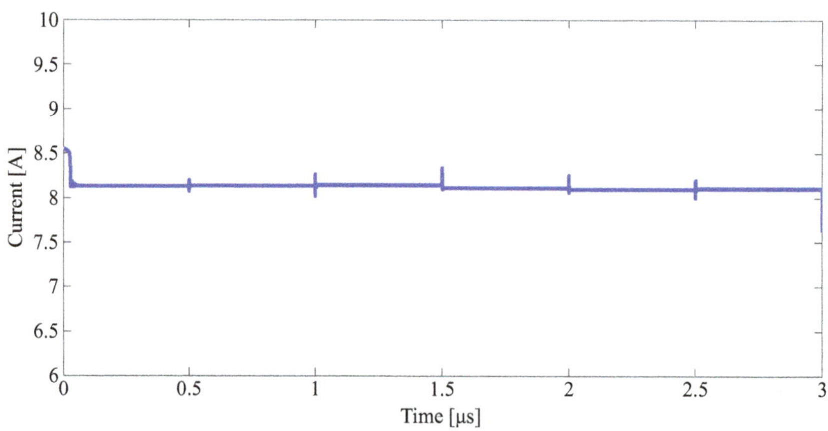

Figure 3.13 PV current variations in response to the temperature changes in scenario 2

oscillations in MPP, which leads to less stress on the components of the PV system. The maximum allowable overshoot of 0.3 can be seen in step-change times, which the controller compensates quickly. The power–voltage curve of the PV system when temperature step-changes are applied to the PV system is represented in Figure 3.14. The AESC has tracked the maximum power points of the PV system precisely and has maintained the maximum point during the system operation. Figure 3.15 depicts that AESC tracks the MPP of the PV system. Then, it regulates the system around this point. Similarly, as seen in Figure 3.16, the tracked output power of the proposed AESC has more minor oscillations in MPP compared with the conventional P&O method. Also, the suggested method yields a faster response than the P&O method while occurring sudden changes in ambient temperature.

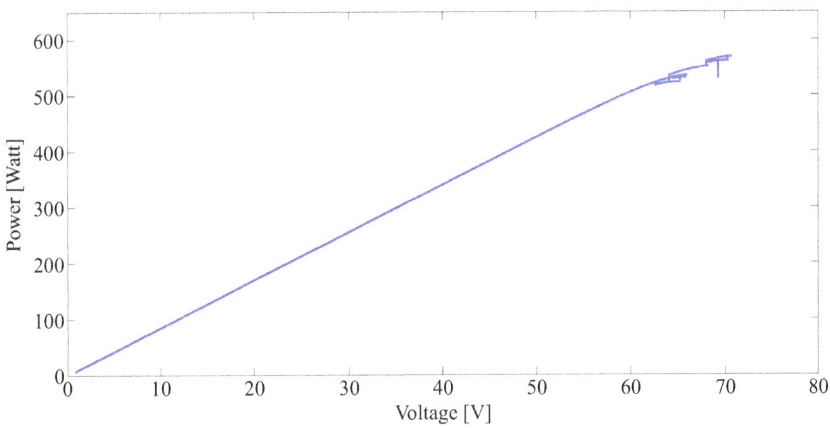

Figure 3.14 Power–voltage characteristic of the PV system in scenario 2

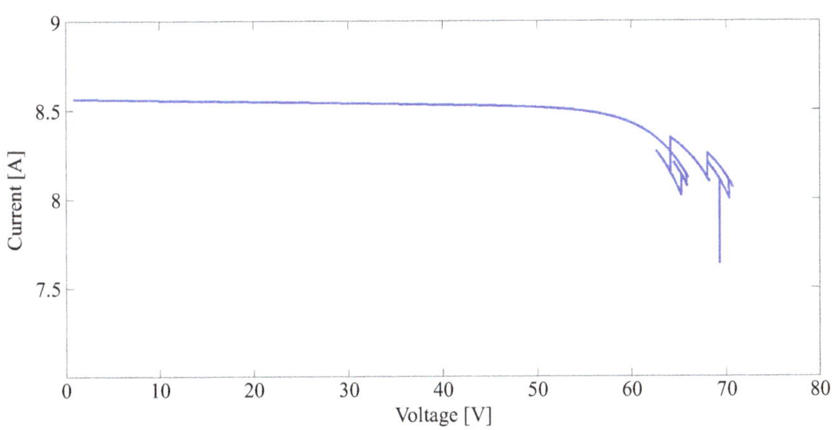

Figure 3.15 Current–voltage curve in scenario 2

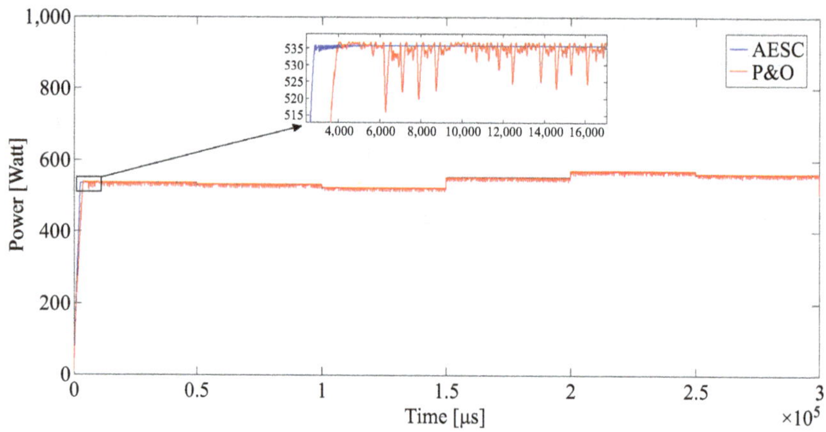

Figure 3.16 The tracked output power of proposed AESC and conventional P&O methods during sudden changes in ambient temperature

3.5 Conclusion

In this chapter, a novel model-free AESC technique has been proposed for the MPPT issue of PV panels in PV systems. The proposed model-free method requires only the power values to obtain the optimal operating points, which accordingly increases the response speed of the suggested method. Meanwhile, a high-gain DC-DC converter was utilized in the considered system with a minimum number of components and high gain per component. The minimum number of components leads to lower power losses and ease of control. The high-gain DC–DC converter used in this chapter is suitable for microgrids, electric vehicles, and medium renewable domestic applications. A 730 W solar module was implemented in MATLAB–Simulink for evaluation of the considered system. Two different scenarios have been considered and simulated to verify the correct performance of the proposed AESC method during dynamic changes in solar irradiation and ambient temperature. The comparative analysis certifies the supremacy of the proposed AESC technique over the conventional P&O technique. Simulation and comparison results confirm that the required response time of the suggested AESC for reaching the MPP is shorter than that of the conventional P&O method. Furthermore, the simulation and comparison results validate that the suggested AESC method provides fewer power oscillations in steady-state response. The AESC method has a power oscillation of about 0.6 W in steady-state response, while the conventional P&O method has an average power oscillation of 8 W. Future works can be conducted on investigating the load variations, grid-connected mode of the considered system and combination with inverters, implementation of other ESC methods and their comparison with the AESC method, and implementation of AESC method on other types of renewable resources.

References

[1] Padmanaban S, Priyadarshi N, Bhaskar MS, *et al.* A hybrid photovoltaic-fuel cell for grid integration with java-based maximum power point tracking: experimental performance evaluation. *IEEE Access*. 2019;7:82978–82990.

[2] Ponnusamy P, Sivaraman P, Almakhles DJ, *et al.* A new multilevel inverter topology with reduced power components for domestic solar PV applications. *IEEE Access*. 2020;8:187483–187497.

[3] Khatoonabad SD, Varesi K, and Padmanaban S. 20. In: *Photovoltaic-Based Switched-Capacitor Multi-Level Inverters with Self-Voltage Balancing and Step-Up Capabilities*. New York, NY: John Wiley & Sons, Ltd; 2020. p. 549–582. Available from: https://onlinelibrary.wiley.com/doi/abs/10.1002/9781119760801.ch20.

[4] Tafti HD, Konstantinou G, Townsend CD, *et al.* Extended functionalities of photovoltaic systems with flexible power point tracking: recent advances. *IEEE Transactions on Power Electronics*. 2020;35(9):9342–9356.

[5] Li X, Wen H, Hu Y, *et al.* A comparative study on photovoltaic MPPT algorithms under EN50530 dynamic test procedure. *IEEE Transactions on Power Electronics*. 2021;36(4):4153–4168.

[6] de Jesus VMR, Cupertino AF, Xavier LS, *et al.* Comparison of MPPT strategies in three-phase photovoltaic inverters applied for harmonic compensation. *IEEE Transactions on Industry Applications*. 2019;55(5):5141–5152.

[7] Deliri S, Varesi K, and Padmanaban S. An extendable single-input reduced-switch 11-level switched-capacitor inverter with quintuple boosting factor. *IET Generation, Transmission & Distribution*. 2022;17(3):621–631. Available from: https://ietresearch.onlinelibrary.wiley.com/doi/abs/10.1049/gtd2.12416.

[8] Manoharan P, Subramaniam U, Babu TS, *et al.* Improved perturb and observation maximum power point tracking technique for solar photovoltaic power generation systems. *IEEE Systems Journal*. 2021;15(2):3024–3035.

[9] Schuss C, Fabritius T, Eichberger B, *et al.* Moving photovoltaic installations: impacts of the sampling rate on maximum power point tracking algorithms. *IEEE Transactions on Instrumentation and Measurement*. 2019;68(5):1485–1493.

[10] Samani L and Mirzaei R. Model predictive control method to achieve maximum power point tracking without additional sensors in stand-alone renewable energy systems. *Optik*. 2019;185:1189–1204. Available from: https://www.sciencedirect.com/science/article/pii/S0030402619305364.

[11] Shams I, Mekhilef S, and Tey KS. Maximum power point tracking using modified butterfly optimization algorithm for partial shading, uniform shading, and fast varying load conditions. *IEEE Transactions on Power Electronics*. 2021;36(5):5569–5581.

[12] Priyadarshi N, Padmanaban S, Kiran Maroti P, *et al.* An extensive practical investigation of FPSO-based MPPT for grid integrated PV system under

variable operating conditions with anti-islanding protection. *IEEE Systems Journal.* 2019;13(2):1861–1871.

[13] Guo S, Abbassi R, Jerbi H, *et al.* Efficient maximum power point tracking for a photovoltaic using hybrid shuffled frog-leaping and pattern search algorithm under changing environmental conditions. *Journal of Cleaner Production.* 2021;297:126573. Available from: https://www.sciencedirect.com/science/article/pii/S0959652621007939.

[14] Li X, Wen H, Hu Y, *et al.* Modified beta algorithm for GMPPT and partial shading detection in photovoltaic systems. *IEEE Transactions on Power Electronics.* 2018;33(3):2172–2186.

[15] Pal RS, and Mukherjee V. A novel population based maximum point tracking algorithm to overcome partial shading issues in solar photovoltaic technology. *Energy Conversion and Management.* 2021;244:114470. Available from: https://www.sciencedirect.com/science/article/pii/S0196890421006464.

[16] Veerapen S, Wen H, Li X, *et al.* A novel global maximum power point tracking algorithm for photovoltaic system with variable perturbation frequency and zero oscillation. *Solar Energy.* 2019;181:345–356. Available from: https://www.sciencedirect.com/science/article/pii/S0038092X19300945.

[17] Li G, Jin Y, Akram MW, *et al.* Application of bio-inspired algorithms in maximum power point tracking for PV systems under partial shading conditions: a review. *Renewable and Sustainable Energy Reviews.* 2018;81:840–873. Available from: https://www.sciencedirect.com/science/article/pii/S1364032117311760.

[18] Issaadi S, Issaadi W, and Khireddine A. New intelligent control strategy by robust neural network algorithm for real time detection of an optimized maximum power tracking control in photovoltaic systems. *Energy.* 2019;187:115881. Available from: https://www.sciencedirect.com/science/article/pii/S0360544219315531.

[19] Ziaei A, Kharrati H, Salim M, *et al.* Offline neural network based-fault tolerant control for vertical tail damaged aircraft. In: *2021 7th International Conference on Control, Instrumentation and Automation (ICCIA)*; 2021. p. 1–5.

[20] Oliveira TR, Krstić M, and Tsubakino D. Extremum seeking for static maps with delays. *IEEE Transactions on Automatic Control.* 2017;62(4):1911–1926.

[21] Salsbury TI, House JM, and Alcala CF. Self-perturbing extremum-seeking controller with adaptive gain. *Control Engineering Practice.* 2020;101:104456. Available from: https://www.sciencedirect.com/science/article/pii/S0967066120300952.

[22] Ziaei A, Kharrati H, Salim M, *et al.* Reinforcement learning-based optimal fault-tolerant control for offshore platforms. *Proceedings of the Institution of Mechanical Engineers, Part I: Journal of Systems and Control Engineering*. 2022;236(6):1187–1196. Available from: https://doi.org/10.1177/09596518221080323.

[23] Ziaei A, Kharrati H, and Rahimi A. Fault-tolerant control for nonlinear offshore steel jacket platforms based on reinforcement learning. *Ocean Engineering.* 2022;246:110247. Available from: https://www.sciencedirect.com/science/article/pii/S0029801821015572.

[24] Liu SJ and Krstic M. Stochastic averaging in discrete time and its applications to extremum seeking. *IEEE Transactions on Automatic Control.* 2016; 61(1):90–102.

[25] Calli B, Caarls W, Jonker P, *et al.* Comparison of extremum seeking control algorithms for robotic applications. In: 2012 *IEEE/RSJ International Conference on Intelligent Robots and Systems*; 2012. p. 3195–3202.

[26] Sadatieh S, Dehghani M, Mohammadi M, *et al.* Extremum-seeking control of left ventricular assist device to maximize the cardiac output and prevent suction. *Chaos, Solitons & Fractals.* 2021;148:111013. Available from: https://www.sciencedirect.com/science/article/pii/S0960077921003672.

[27] Sotiropoulos FE and Asada HH. A model-free extremum-seeking approach to autonomous excavator control based on output power maximization. *IEEE Robotics and Automation Letters.* 2019;4(2):1005–1012.

[28] Kumar S, Zwall MR, Bolívar-Nieto EA, *et al.* Extremum seeking control for stiffness auto-tuning of a quasi-passive ankle exoskeleton. *IEEE Robotics and Automation Letters.* 2020;5(3):4604–4611.

[29] Zhou D, Al-Durra A, Matraji I, *et al.* Online energy management strategy of fuel cell hybrid electric vehicles: a fractional-order extremum seeking method. *IEEE Transactions on Industrial Electronics.* 2018;65(8):6787–6799.

[30] Vandermeulen I, Guay M, and McLellan PJ. Distributed control of high-altitude balloon formation by extremum-seeking control. *IEEE Transactions on Control Systems Technology.* 2018;26(3):857–873.

[31] Xu Q and Cai L. Active braking control of electric vehicles to achieve maximal friction based on fast extremum-seeking and reachability. *IEEE Transactions on Vehicular Technology.* 2020;69(12):14869–14883.

[32] Sharma S and Alleyne AG. Extremum seeking control of battery powered vapor compression systems for commercial vehicles. *International Journal of Refrigeration.* 2020;115:63–72. Available from: https://www.sciencedirect.com/science/article/pii/S0140700720301006.

[33] Maroti PK, Padmanaban S, Holm-Nielsen JB, *et al.* A new structure of high voltage gain SEPIC converter for renewable energy applications. *IEEE Access.* 2019;7:89857–89868.

[34] Zhang F, Thanapalan K, Procter A, *et al.* Adaptive hybrid maximum power point tracking method for a photovoltaic system. *IEEE Transactions on Energy Conversion.* 2013;28(2):353–360.

[35] Varesi K, Hassanpour N and Saeidabadi S. Novel high step-up DC–DC converter with increased voltage gain per devices and continuous input current suitable for DC microgrid applications. *International Journal of Circuit Theory and Applications.* 2020;48(10):1820–1837. Available from: https://onlinelibrary.wiley.com/doi/abs/10.1002/cta.2804.

[36] Zhang C and Ordonez R. Numerical optimization-based extremum seeking control of LTI systems. In: *Proceedings of the 44th IEEE Conference on Decision and Control*; 2005. p. 4428–4433.

[37] Zhang C and Ordonez R. Non-gradient extremum seeking control of feedback linearizable systems with application to ABS design. In: *Proceedings of the 45th IEEE Conference on Decision and Control*; 2006. p. 6666–6671.

[38] Chen CT. Linear system theory and design. In *The Oxford Series in Electrical and Computer Engineering.* Oxford: Oxford University Press; 2014. Available from: https://books.google.com/books?id=XyPAoAEACAAJ.

[39] Zhang C and Ordóñez R. Robust and adaptive design of numerical optimization-based extremum seeking control. *Automatica.* 2009;45(3):634–646. Available from: https://www.sciencedirect.com/science/article/pii/S0005109808004895.

[40] Zhang C and Ordonez R. Numerical optimization-based extremum seeking control with application to ABS design. *IEEE Transactions on Automatic Control.* 2007;52(3):454–467.

[41] Nocedal J and Wright SJ. *Numerical Optimization*, 2nd ed. New York, NY: Springer; 2006.

[42] Bertsekas D. Nonlinear programming. In *Athena Scientific Optimization and Computation Series.* Athena Scientific; 2016. Available from: https://books.google.com/books?id=TwOujgEACAAJ.

[43] Kolda TG, Lewis RM and Torczon V. Optimization by direct search: new perspectives on some classical and modern methods. *SIAM Review.* 2003;45(3):385–482. Available from: https://doi.org/10.1137/S003614450242889.

[44] Boyd S and Vandenberghe L. *Convex Optimization.* Cambridge, MA: Cambridge University Press; 2004. Available from: http://www.amazon.com/exec/obidos/redirect?tag=citeulike-20&path=ASIN/0521833787.

[45] Rezk H and Eltamaly AM. A comprehensive comparison of different MPPT techniques for photovoltaic systems. *Solar Energy.* 2015;112:1–11. Available from: https://www.sciencedirect.com/science/article/pii/S0038092X14005428.

Chapter 4

A control scheme to optimize efficiency of GaN-based DC–DC converters

Amit Kumar Singha[1]

Half-bridge switch cells are frequently employed in DC–DC converter circuits to improve efficiency. However, under light-load conditions, employing both switches degrades efficiency as the switching and driver losses become dominant. In metal-oxide–semiconductor field-effect transistor (MOSFET)-based converters, the sync field-effect transistor (FET) is turned OFF to eliminate these losses and during this duration sync FET's body diode enables reverse conduction. In a gallium nitride (GaN) FET, reverse conduction is possible through its 2-D electron gas (2DEG) channel when the voltage across source-to-drain is higher than the gate threshold voltage. Unlike MOSFET's body diode, it has no reverse recovery loss. However, the voltage drop across source-to-drain during reverse conduction is significantly higher than forward drop of the MOSFET's body diode. The voltage drop can be reduced using a Schottky diode in anti-parallel configuration with the sync FET. MOSFET-based converters also use zero current detection (ZCD) circuitry to enable discontinuous conduction mode (DCM) of operation using the sync FET. Both the above-mentioned techniques use additional components to realize DCM; therefore, the techniques are not cost effective. To optimize the efficiency under light load conditions, this chapter proposes a novel control scheme that can emulate the discontinuous conduction mode (DCM) of operation without using a Schottky diode or a ZCD circuitry. Furthermore, the proposed control technique is simple to implement and it can be easily combined with the existing light-load control techniques. A prototype of the GaN boost converter is designed and a field-programmable gate array (FPGA) device is used to implement the proposed scheme.

4.1 Introduction

Switching DC–DC converters employing half-bridge switch cells [1] demonstrate significant efficiency improvement in high-load conditions compared to linear

[1]SCEE, IIT Mandi, India

regulators [2] and switched-capacitor-based converters [3]. However, under light-load conditions, half-bridge switch cells degrade efficiency as the switching and driver losses of the sync FET become dominant. Thus, to eliminate these losses in MOSFET-based converters, the sync FET is turned off and the body diode or an antiparallel Schottky diode is used for reverse conduction. In recent years, GaN FETs [4] have shown a significant efficiency improvement in the field of high-frequency power electronics despite its gate driver design challenges [5]. Like MOSFETs, GaN half-bridge switch cells are now frequently employed in various power electronic converters. To achieve DCM operation in MOSFET-based converters, body diode of the sync FET is enabled for reverse conduction. In a GaN FET, reverse conduction is possible through its 2-D electron gas (2DEG) channel (see Figure 4.1) when the voltage across source-to-drain is higher than V_{th} (gate threshold voltage) [6]. Unlike MOSFET's body diode, it has no reverse recovery loss. However, the voltage across source-to-drain in a GaN FET during reverse conduction is much higher than a Si-MOSFET's body diode and it can be computed by the following expression [7]:

$$V_{sd} \approx -(V_{th} + i_r R_{ds,on}), \tag{4.1}$$

where i_r is the reverse current and $R_{ds,on}$ is the on-resistance of the channel. Power loss P_{dt} during reverse conduction can be calculated using the following expression:

$$P_{dt} = V_{sd} i_r t_d F_s. \tag{4.2}$$

Here, t_d is the time duration of reverse conduction and F_s represents the converter's switching frequency. This loss can be minimized if V_{sd} can be reduced. The voltage drop can be reduced by placing an antiparallel Schottky diode with the sync GaN FET. However, this would increase the cost of implementation.

Recently, digital controllers have gained huge popularity in the power management industry. This is due to their ease of implementation of advanced control techniques and reprogrammability [8–10]. Different digital controllers were proposed to realize the popular current mode and voltage mode control techniques. Light load control techniques for efficiency improvement in MOSFET-based converters were proposed recently [11,12]. However, the full potential of digital controllers is yet to be exploited.

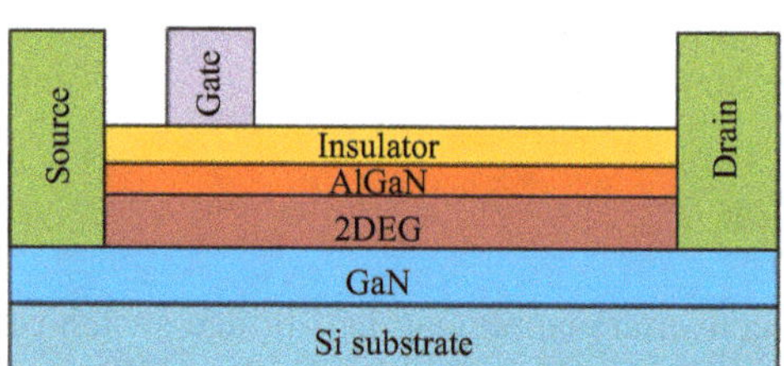

Figure 4.1 Lateral structure of GaN FET. Reverse conduction is possible through the 2DEG channel between GaN and AlGaN.

Light load efficiency improvement is needed for battery-operated portable devices to extend their battery life. Various light load efficiency improvement techniques were reported in the literature [13–21]. Phase shedding schemes were proposed in [13,14] to improve the light load efficiency of a multiphase buck converter. Efficiency can be optimized in the single-phase buck converter by employing pulse skipping modulation (PSM) [15,16,22]. Driver loss is significant in MOSFET-based converters operating under light load conditions. Driver loss minimization techniques were reported in [17,18]. By varying the gate driver voltage, loss of the buck converter was reduced in [18]. Switching frequency of the converter affects the switching loss and driver loss. Pulse frequency modulation (PFM) [19,20,23] techniques were proposed to adjust the frequency based on the load to optimize the light load efficiency. Light load loss of the dual active bridge was optimized in [21] using a digital platform based compensation method. Using dynamic width control and sigma-delta PSM technique, light load efficiency of the buck converter was improved [24]. An adaptive peak current control technique was proposed in [25] to optimize the efficiency of the single-inductor multi-output (SIMO) converter. Efficiency of the converters can be improved by employing intermediate energy storage and periodically turning off the converter [26]. Modulating and recycling of gate charge can also help to improve the light load efficiency [17]. The above-mentioned techniques are suitable for MOSFET-based converters. GaN-based converters pose different challenges as the reverse conduction loss is higher though the driver loss is low due to smaller gate-to-source capacitance. Therefore, to improve the efficiency of GaN-based converters under light-load conditions, a suitable control technique is required which can exploit the benefits and eliminate the drawbacks of GaN devices.

The efficiency of PSM can be improved further by turning ON the synchronous FET. Synchronous FET turning ON time can be obtained using zero current detection (ZCD) circuitry and input voltage sensing [23,24,27–29]. The ZCD circuit uses switch-node voltage and in switching DC–DC converters the switch-node voltage is noisy. Thus, determining exact turn ON time of the synchronous FET would be difficult. This chapter proposes a control technique which is robust as it does not rely on switch-node voltage ZCD circuitry. However, the proposed technique needs to load current and input voltage information which are already available in battery operated portable applications [11,19]. Thus, the proposed method does not need any additional circuitry. Furthermore, the proposed scheme can be implemented in the digital domain. Therefore, it will be cost effective due to the ever decreasing cost of digital platform. The proposed method is flexible and it can be easily combined with the existing light load efficiency improvement schemes such as PSM and PFM. The effectiveness and flexibility of the proposed method are judged with an experimental setup. An FPGA platform is used to implement the proposed scheme and a GaN boost converter is developed using GS61008P from GaN systems. Experimental results demonstrate that the proposed scheme is more effective compared to the conventional asynchronous operation achieved using an antiparallel Schottky diode with the sync FET and conventional PSM.

Organization of the chapter is as follows. Section 4.2 presents the proposed control scheme. Experimental verification is carried out in Section 4.3. Section 4.4 concludes the chapter.

4.2 Proposed control scheme

A schematic of the synchronous boost converter using GaN FETs is shown in Figure 4.2. Here, v_{in} and v_o are the input and output voltages, respectively. R, L, and C_o refer to the load resistance, inductor, and output capacitor, respectively. The two FETs operate in a complementary fashion. Thus, to eliminate shoot-through, dead-time is applied between the state transition of two FETs. During dead-time, if the inductor current i_L is positive, the high-side FET enables reverse conduction. The reverse conduction is modeled as anti-parallel body diode as shown in Figure 4.2. The inductor current waveform under DCM is shown in Figure 4.3(a). During d_1T time duration, switch S_L is ON (see Figure 4.3(b)) and i_L is rising. During d_2T interval, switch S_L is OFF; thus, i_L flows through the anti-parallel body diode of S_H as shown in Figure 4.3(c). At the end of d_2T duration, the inductor current reaches zero. During d_3T interval (see Figure 4.3(a)), the inductor current stays at zero as the high-side switch is in OFF state as shown in Figure 4.3(d). The inductor current waveform shown in Figure 4.3(a) is possible to emulate in a synchronous boost converter if the high-side switch S_H is turned OFF when the inductor current reaches zero (after d_2T duration). This is possible to achieve with a current sensor. However, this would increase the cost of implementation. This chapter proposes a digital control technique which does not need a current sensor. The proposed method is discussed below.

Under discontinuous conduction mode of operation, the voltage gain of the boost converter is as follows:

$$\frac{v_o}{v_{in}} = \frac{d_1 + d_2}{d_2}. \tag{4.3}$$

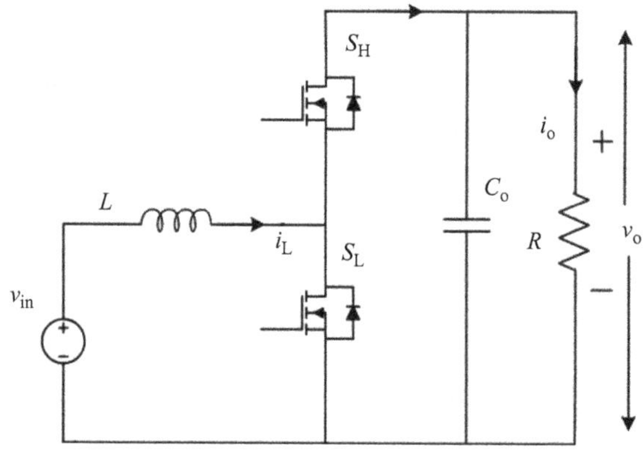

Figure 4.2 Schematic of a synchronous boost converter

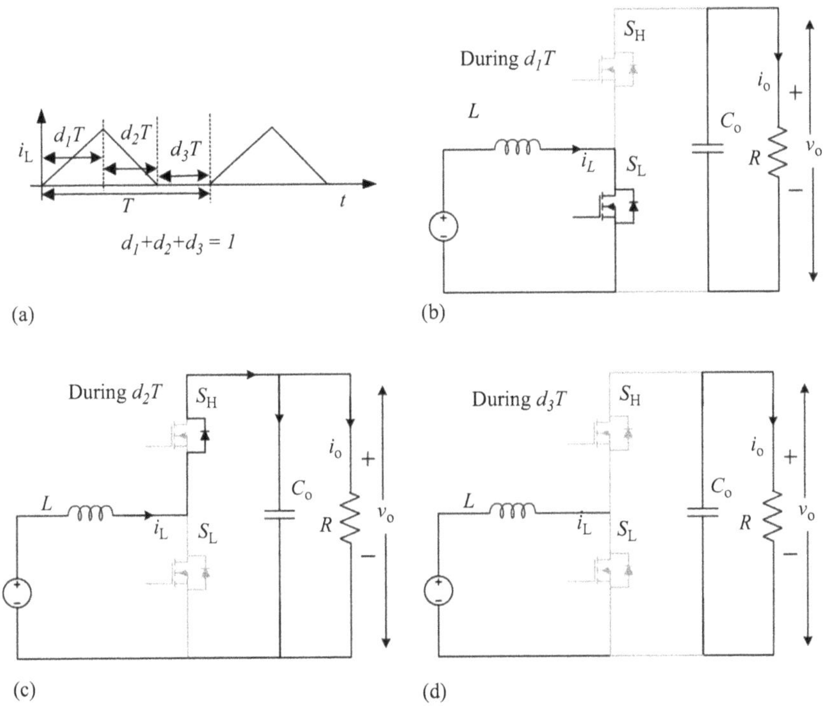

Figure 4.3 (a) The inductor current waveform under DCM. (b) Low-side switch is conducting. (c) High-side switch is turned OFF and the current is flowing through its body diode. (d) Both the switches are OFF and sync FET's body diode is in reverse bias; thus, the inductor current stays at zero.

The output current i_o can be expressed as

$$i_o = \frac{v_o}{R} = \frac{v_{in} d_1 d_2 T}{2L}. \tag{4.4}$$

From (4.4), d_2 expression can be derived as

$$d_2 = \frac{2Lv_o}{Rv_{in}Td_1}. \tag{4.5}$$

To compute d_2, d_1 is required. An illustrative block diagram of the proposed scheme is shown in Figure 4.4 where $u_{c,L} = d_1$ is computed by comparing the output of the voltage loop controller $G_c(z)$ with a sawtooth wave V_{sw1} using the comparator Comp A:

$$d_1 = \frac{v_c[n]}{v_p}. \tag{4.6}$$

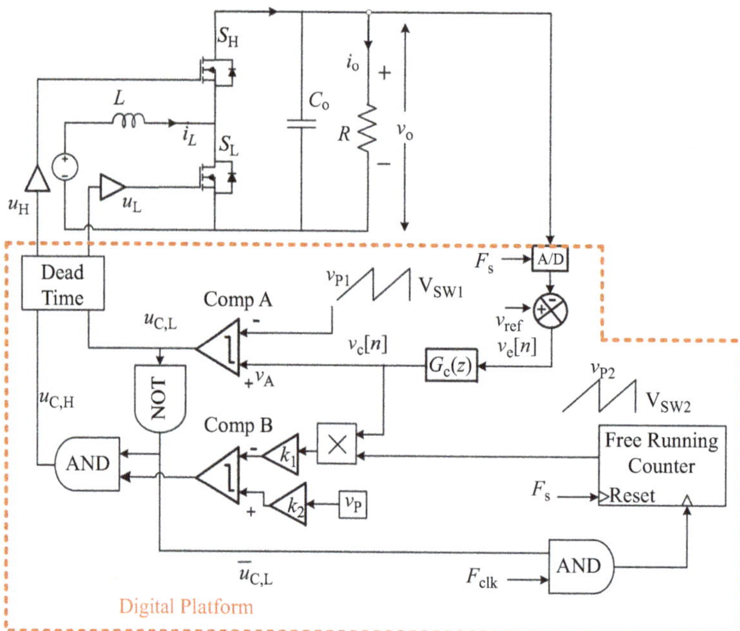

Figure 4.4 Proposed control scheme

where $v_c[n]$ is the output of the digital controller $G_c(z)$ and $v_{p1} = v_p$ is the peak value of the sawtooth wave. The output of the comparator is passed through a dead time block to produce u_L. From (4.5) and (4.6), d_2 can be expressed as:

$$d_2 = \frac{k_2 v_p}{k_1 v_c[n]}. \tag{4.7}$$

where $k_1 = Rv_{in}T$ and $k_2 = 2Lv_o$. Based on the above expression, d_2 can be obtained if v_p is multiplied with k_2 and compared with a sawtooth wave whose peak value is $k_1 v_c[n]$.

To produce d_2, a free running counter is used which gets reset at the rising edge of the converter clock F_s (see Figure 4.4). Illustrative waveforms of the proposed scheme are shown in Figure 4.5. The counter starts counting when it starts receiving a high-frequency clock F_{clk}. To produce d_2 at right time instant, F_{clk} is multiplied with $\bar{u}_{C,L}$. The signal $\bar{u}_{C,L}$ is an inversion of $u_{C,L}$. Output of the free running counter is a sawtooth V_{sw2} which is multiplied with $v_c[n]$ and k_1 and compared with $k_2 v_p$ using the comparator Comp B. Peak value of the sawtooth wave $v_{p2} = 1$. Output of the comparator is ANDed with $\bar{u}_{C,L}$ to produce $u_{C,H}$ which is equivalent to d_2. A dead time block is used to insert a dead time between $u_{C,L}$ and $u_{C,H}$ to avoid shoot-through. u_H is the gate signal which goes to the high-side switch S_H.

Gains k_1 and k_2 depend on system parameters. The output voltage is already sampled; however, the input voltage needs to be sampled and the load information needs to be provided to implement the proposed technique in real-time.

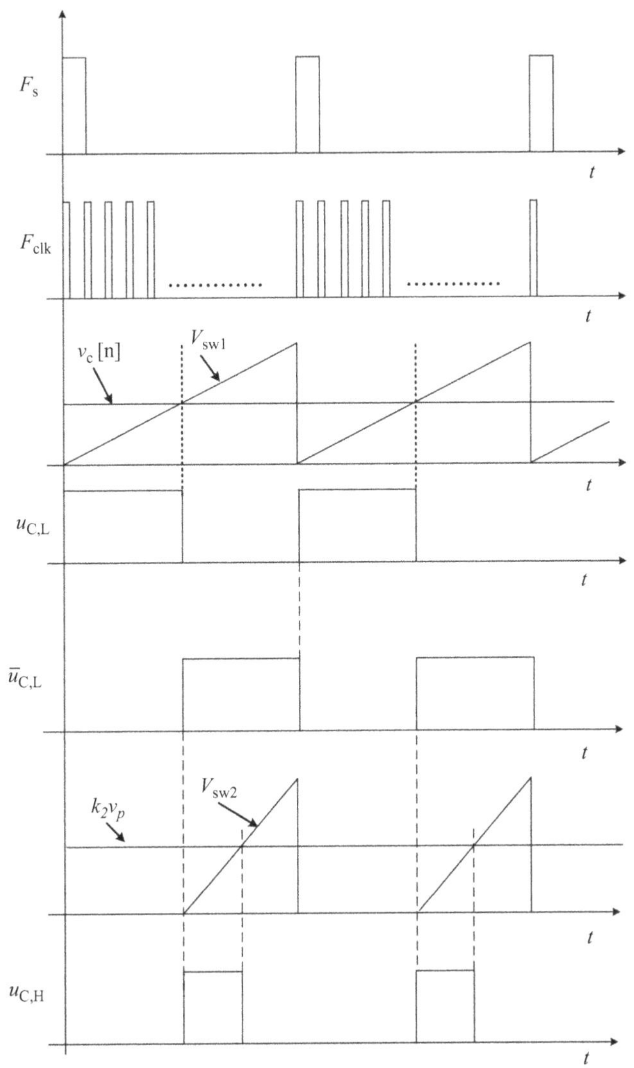

Figure 4.5 Illustrative waveforms of the proposed control scheme. F_s is the converter clock and F_{clk} is the high-frequency clock. Free running counter starts counting when it receives F_{clk}.

4.3 Simulation and experimental verification

A MATLAB®/Simulink® model of the synchronous boost converter is prepared to validate the proposed control scheme in simulation environment. The following parameter set is used for the simulation: $L = 15\ \mu$H, $v_{in} = 140$ V, $v_o = 240$ V, $R = 55\ \Omega$, switching frequency $F_s = 200$ kHz, and $C_o = 600\ \mu$F. Figure 4.6 shows the simulated inductor current waveform at a power level of 1 kW. The gate signal u_L is produced using the voltage controller $G_c(z)$ loop and the gate signal u_H is generated using the proposed control scheme. To emulate DCM, the gate signal u_H goes zero as the inductor current reaches zero. Therefore, the simulation validates that the proposed scheme can produce gate signals for both the FETs in such a way that DCM can be achieved without using a Schottky diode. The simulation cannot accurately capture the effect of the proposed scheme on the converter efficiency; thus, a GaN-based boost converter prototype is built to validate the proposed control strategy experimentally. A low power setup is built as the chosen GaN FETs are suitable for low voltage applications. However, the proposed scheme can be validated even with high-voltage GaN FETs as well.

Both the GaN FETs are realized using GS61008P from GaN systems and half-bridge driver LM5113QDPRRQ1 from TI is employed to drive these switches. The parameter set chosen for the prototype is given in Table 4.1. The proposed control scheme is implemented using a Spartan-7 FPGA board (Arty S7) from Digilent. Spartan-7 FPGA has a dual-channel 12-bit ADC which is used to sample the output voltage v_o of the boost converter. To implement conventional DCM operation with a

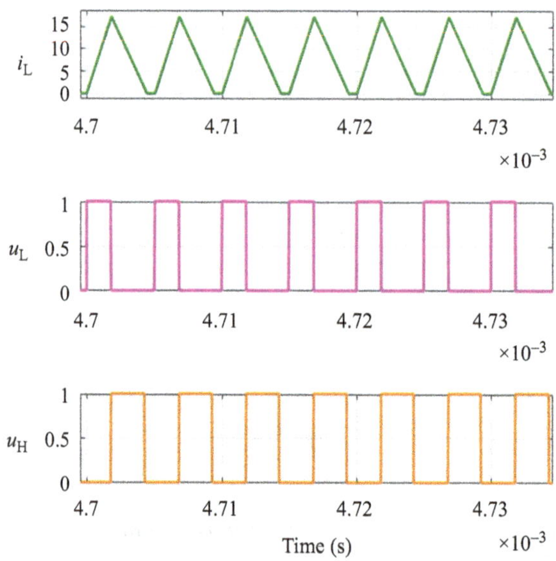

Figure 4.6 Discontinuous conduction mode operation is achieved using the proposed control strategy at 1 kW

Table 4.1 Parameter set of the GaN boost converter

v_{in} (V)	v_o (V)	L (μH)	C_o (μF)	R (Ω)	F_s
14	24	3.3	288	[10–50]	200 kHz

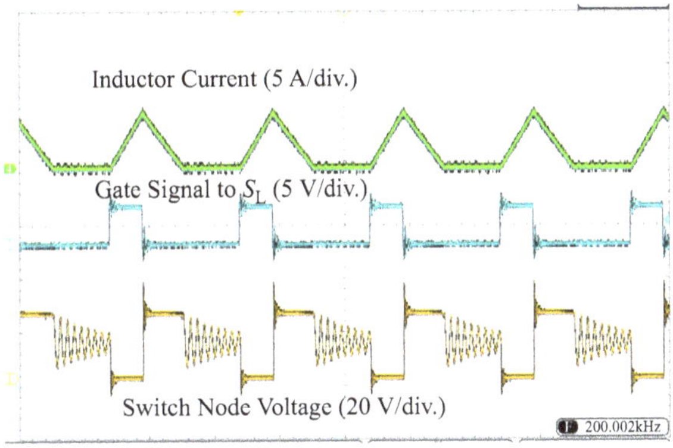

Figure 4.7 Discontinuous conduction mode operation is achieved using a Schottky diode. Green trace shows the inductor current (5 A/div.), gate signal to the low-side FET S_L *is shown in cyan trace (5 V/div.), and yellow trace depicts the switch-node voltage (20 V/div.).*

diode, the high-side GaN FET is replaced with a Schottky diode (SS3H10-E3/9AT from VISHAY). Figure 4.7 shows the waveforms of the boost converter when a Schottky diode is used to achieve DCM. The input voltage is set at 14 V and the load resistance is chosen as 50 Ω. The inductor current is shown in green trace (5 A/div.). Gate signal to the low side FET S_L is shown in cyan trace (5 V/div.) and switch node voltage is shown in yellow trace (20 V/div.). The switch node voltage shows oscillations when the inductor current becomes zero. Similar behavior can also be achieved in the synchronous boost converter using the proposed scheme by controlling the ON time of the sync FET. Figure 4.8 shows that two gate signals are applied to the high-side and low-side FETs. Gate signal for the low-side FET (shown in magenta trace) is coming from the upper control loop which contains the voltage loop controller. The gate signal for the high-side FET is generated using the proposed control scheme and shown in cyan trace. It can be seen that the gate signal for the high-side FET goes low when the current becomes zero. Thus, the converter replicates DCM using an active switch without using a ZCD circuit or diode. The proposed scheme is further tested with a power of 55 W ($R = 10.5$ Ω) and shown in Figure 4.9. The input voltage and other parameters remain the same. At $R = 10.5$ Ω, the converter's output load has increased; therefore, d_2T duration has to be increased. The proposed control scheme adjusts the ON time duration of the high-side FET by adjusting k_1.

Effectiveness of the proposed control scheme is judged from the efficiency point of view. Efficiency of the conventional DCM with Schottky diode and efficiency obtained using the proposed control scheme are plotted by varying the power from 14 W to 22 W and shown in Figure 4.10. The orange colour trace is obtained using the proposed control technique and the blue colour trace is obtained using a Schottky diode. The efficiency results are taken considering the driver loss. The driver loss has been calculated using the method discussed in [30]. The input

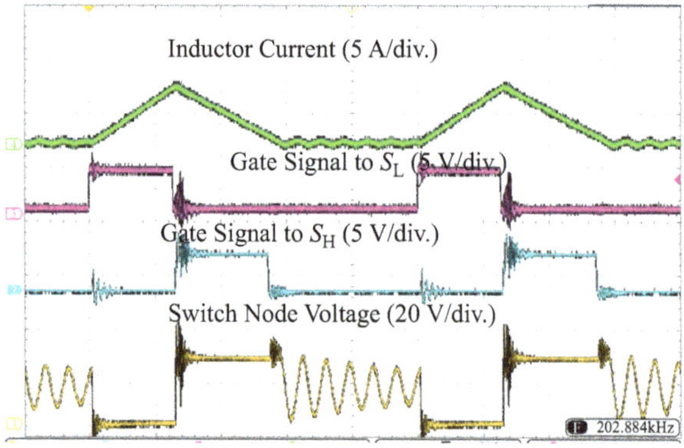

Figure 4.8 Discontinuous conduction mode operation is achieved using the proposed control strategy. The green trace shows the inductor current (5 A/div.), gate signal to S_L is shown in magenta trace (5 V/div.), gate signal to S_H is produced using the proposed scheme and shown in cyan trace (5 V/div.), and yellow trace depicts the switch-node voltage (20 V/div.).

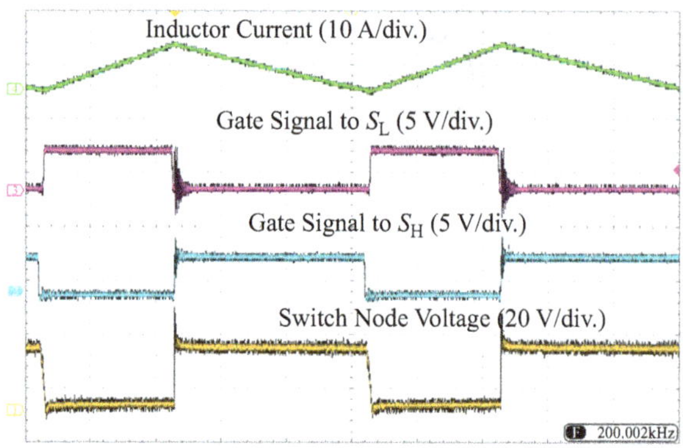

Figure 4.9 Discontinuous conduction mode operation is achieved using the proposed control strategy with $R = 10.5\ \Omega$

voltage of the driver is measured with a $5\frac{1}{2}$ digit multimeter. The total gate charge value is taken from the datasheet. The driving loss in a GaN FET is negligible due to very low gate charge compared to a MOSFET device. Figure 4.10 shows that at 14 W, the proposed scheme provides more than 1% improvement in efficiency. However, the improvement degrades as the power level increases. This indicates that near the boundary of continuous conduction mode (CCM) and DCM, conventional Schottky diode and the proposed schemes have similar efficiency. Thus, it is evident from the plot that the proposed scheme is much more efficient when the converter is in deep DCM than the conventional DCM scheme obtained using Schottky diode.

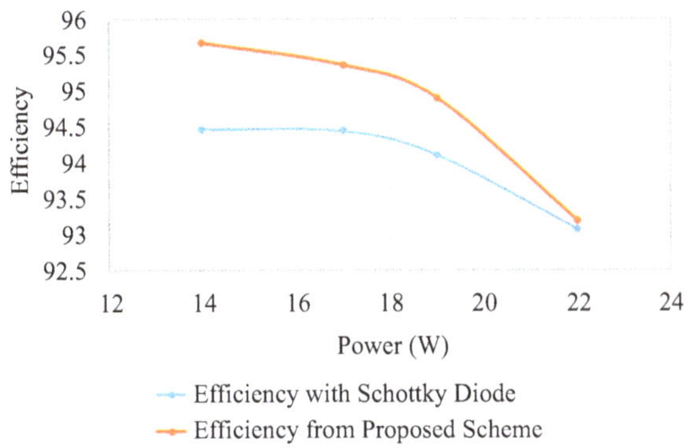

Figure 4.10 Efficiency plot. Blue line is obtained with Schottky diode and orange line is obtained from the proposed scheme.

Table 4.2 Efficiency comparison

Input power (W)	Proposed + PFM (%)	Conventional PFM (%)
13	97.43	84.93
15	97.62	84.43
18	97.62	86.2

One of the advantages of the proposed scheme is that it can be combined with the existing light load efficiency improvement techniques such as PFM or PSM to optimize the converter's efficiency further. Duty for the low-side FET will come from the upper loop. This can be converted to a digital number using a time-to-digital converter [31]. This digital number can be used to calculate the duty for the sync FET using (4.5). To demonstrate the effectiveness and flexibility of the proposed technique, conventional PFM scheme and the proposed scheme with PFM are applied in the in-house GaN-based boost converter. In conventional PFM scheme, the high-side FET is turned OFF whereas sync FET is used when the proposed scheme is combined with PFM. The power is increased from 13 W to 18 W at $v_{in} = 14$ V and experimental efficiency is measured and shown in Table 4.2. At 13 W, the proposed scheme improves the efficiency by 13% compared to conventional PFM scheme. As the power level increases, the proposed scheme maintains the efficiency at 97%. Thus, comparison shows that the proposed scheme can optimize the efficiency of conventional PFM scheme further.

4.4 Conclusions

Reverse conduction in GaN FETs introduces a significant voltage drop between the source and the drain. This drop leads to a significant efficiency degradation in GaN-based converters under light-load conditions. To mitigate this drawback, this chapter has proposed a novel control technique to implement DCM in a synchronous boost converter. The proposed technique does not need a Schottky diode or a current sensor. Furthermore, the proposed method can be easily combined with other existing light load control techniques. Experimental results have demonstrated that the proposed method can optimize the efficiency compared to an asynchronous boost converter. The proposed control technique can be used in other GaN-based power electronic converters.

References

[1] Kohlhepp B, Kübrich D, and Dürbaum T. Switching loss measurement – a thermal approach applied to GaN-half-bridge configuration. In: *2021 23rd European Conference on Power Electronics and Applications (EPE'21 ECCE Europe)*. IEEE; 2021. p. 1–10.

[2] Perez R. A comparative assessment between linear and switching power supplies in portable electronic devices. In: *IEEE International Symposium on Electromagnetic Compatibility*. Symposium Record (Cat. no. 00CH37016). vol. 2. IEEE; 2000. p. 839–843.

[3] Chung H, Brian O, and Ioinovici A. Switched-capacitor-based DC-to-DC converter with improved input current waveform. In: *1996 IEEE International Symposium on Circuits and Systems. Circuits and Systems Connecting the World*. ISCAS 96. vol. 1. IEEE; 1996. p. 541–544.

[4] Runton DW, Trabert B, Shealy JB, *et al.* History of GaN: high-power RF gallium nitride (GaN) from infancy to manufacturable process and beyond. *IEEE Microwave Magazine*. 2013;14(3):82–93.

[5] Bindra A. Recent advances in gate driver integrated circuits for wide-bandgap FETs: manufacturers introduce gate drivers for squeezing the best performance out of WBG devices. *IEEE Power Electronics Magazine*. 2019;6(3):32–38.

[6] Lidow A, De Rooij M, Strydom J, *et al. GaN Transistors for Efficient Power Conversion*. New York, NY: John Wiley & Sons; 2019.

[7] Asad M, Singha AK, and Rao RMS. Dead time optimization in a GaN-based buck converter. *IEEE Transactions on Power Electronics*. 2021;37(3): 2830–2844.

[8] Hagen M and Yousefzadeh V. Applying digital technology to PWM control-loop designs. In: *Power Supply Design Seminar SEM-1800*. vol. 2009; 2008. p. 7–1.

[9] Liu YF, Meyer E, and Liu X. Recent developments in digital control strategies for DC/DC switching power converters. *IEEE Transactions on Power Electronics*. 2009;24(11):2567–2577.

[10] Asad M and Singha AK. Real-time dead-time optimization in a GaN-based boost converter using a digital controller. In: *2021 IEEE 12th Annual Ubiquitous Computing, Electronics & Mobile Communication Conference (UEMCON)*. IEEE; 2021. p. 0723–0729.

[11] Kapat S. Configurable multimode digital control for light load DC–DC converters with improved spectrum and smooth transition. *IEEE Transactions on Power Electronics.* 2016;31(3):2680–2688.
[12] Trescases O and Wen Y. A survey of light-load efficiency improvement techniques for low-power dc-dc converters. In: *8th International Conference on Power Electronics-ECCE Asia*. IEEE; 2011. p. 326–333.
[13] Ahn Y, Jeon I, and Roh J. A multiphase buck converter with a rotating phase-shedding scheme for efficient light-load control. *IEEE Journal of Solid-State Circuits.* 2014;49(11):2673–2683.
[14] Su JT and Liu CW. A novel phase-shedding control scheme for improved light load efficiency of multiphase interleaved DC–DC converters. *IEEE Transactions on Power Electronics.* 2012;28(10):4742–4752.
[15] Luo P, Luo L, Li Z, *et al.* Skip cycle modulation in switching DC-DC converter. In: *IEEE 2002 International Conference on Communications, Circuits and Systems and West Sino Expositions*. vol. 2. IEEE; 2002. p. 1716–1719.
[16] Kapat S, Mandi BC, and Patra A. Voltage-mode digital pulse skipping control of a DC-DC converter with stable periodic behavior and improved light-load efficiency. *IEEE Transactions on Power Electronics.* 2016;31 (4):3372–3379.
[17] Mulligan MD, Broach B, and Lee TH. A 3 MHz low-voltage buck converter with improved light load efficiency. In: *2007 IEEE International Solid-State Circuits Conference. Digest of Technical Papers*. IEEE; 2007. p. 528–620.
[18] Mulligan MD, Broach B, and Lee TH. A constant-frequency method for improving light-load efficiency in synchronous buck converters. *IEEE Power Electronics Letters.* 2005;3(1):24–29.
[19] Sun J, Xu M, Ren Y, *et al.* Light-load efficiency improvement for buck voltage regulators. *IEEE Transactions on Power Electronics.* 2009; 24 (3):742–751.
[20] Zhang X and Maksimovic D. Multimode digital controller for synchronous buck converters operating over wide ranges of input voltages and load currents. *IEEE Transactions on Power Electronics.* 2010;25(8):1958–1965.
[21] Hirose T, Takasaki M, and Ishizuka Y. A power efficiency improvement technique for a bidirectional dual active bridge DC–DC converter at light load. *IEEE Transactions on Industry Applications.* 2014;50(6):4047–4055.
[22] Angkititrakul S and Hu H. Design and analysis of buck converter with pulse-skipping modulation. In: *2008 IEEE Power Electronics Specialists Conference*. IEEE; 2008. p. 1151–1156.
[23] Sahu B and Rincon-Mora GA. An accurate, low-voltage, CMOS switching power supply with adaptive on-time pulse-frequency modulation (PFM) control. *IEEE Transactions on Circuits and Systems I: Regular Papers.* 2007;54(2):312–321.
[24] Yan W, Li W, and Liu R. A noise-shaped buck DC–DC converter with improved light-load efficiency and fast transient response. *IEEE Transactions on Power Electronics.* 2011;26(12):3908–3924.

[25] Huang MH and Chen KH. Single-inductor multi-output (SIMO) DC-DC converters with high light-load efficiency and minimized cross-regulation for portable devices. *IEEE Journal of Solid-State Circuits*. 2009;44(4):1099–1111.

[26] Jang Y and Jovanovic MM. Light-load efficiency optimization method. *IEEE Transactions on Power Electronics*. 2009;25(1):67–74.

[27] Fu W, Tan ST, Radhakrishnan M, *et al.* A DCM-only buck regulator with hysteretic-assisted adaptive minimum-on-time control for low-power microcontrollers. *IEEE Transactions on Power Electronics*. 2015;31 (1):418–429.

[28] Zhou X, Donati M, Amoroso L, *et al.* Improved light-load efficiency for synchronous rectifier voltage regulator module. *IEEE Transactions on Power Electronics*. 2000;15(5):826–834.

[29] Gao Y, Wang S, Li H, *et al.* A novel zero-current-detector for DCM operation in synchronous converter. In: *2012 IEEE International Symposium on Industrial Electronics*. IEEE; 2012. p. 99–104.

[30] Lakkas G. MOSFET power losses and how they affect power-supply efficiency. *Analog Applications*. 2016;10(2016):22–26.

[31] Song J, An Q, and Liu S. A high-resolution time-to-digital converter implemented in field-programmable-gate-arrays. *IEEE Transactions on Nuclear Science*. 2006;53(1):236–241.

Chapter 5

Control design of grid-connected three-phase inverters

Kirti Gupta[1] and Bijaya Ketan Panigrahi[1]

Recently, there is a rapid growth in the deployment of both high and medium power converters to interconnect renewable energy resources to the network. These inverter-interfaced energy resources (IIERs) provide clean and green production of energy, which can be either connected to the grid or can operate in off-grid mode [1]. As the operating challenges related to intermittent power generation through these renewable sources of energy (like solar, wind, etc.) can be overcome by interconnecting these sources to the grid, hence this chapter elaborates the intelligent control technique of these inverters. A brief overview of various inverter topologies along with a detailed study of the control architecture of grid-connected inverters is presented. An implementation of the control scheme on two different testbeds is demonstrated. The first is the *real-time (RT) co-simulation testbed* and the second is the *power hardware-in-loop testbed* (PHIL). A test case for each of the testbeds is presented to demonstrate the ability of these testbeds.

5.1 Introduction

To aim decarbonization goals, there is widespread adoption of renewable energy sources like solar, wind, hydro, etc. [2–4]. These are interconnected to the AC system through the medium and high power converters according to the system voltage level. Recently there is a huge surge in the integration of these IIERs in the electrical infrastructure. Due to the large penetration of these renewable energy resources integrated with power-electronic devices, there is a decrease in system inertia which can cause frequency instability and reliability issues [5,6]. However, to combat these issues various intelligent control strategies proposed by the researchers are incorporated. Broadly there are two configurations available for IIERs namely, (i) grid-connected mode and (ii) off-grid mode [7]. The inverters integrating the renewable energy sources into the grid are generally termed as grid-following (GFL) inverters, whereas the inverters running in autonomous mode are

[1]Department of Electrical Engineering, Indian Institute of Technology, India

known as grid forming (GFM) inverters [8,9]. Both these types of inverters have their own functionality and control architecture. The inverters can also be equipped with dual functionality i.e., to operate in both grid and off-grid modes, by modifying the control strategy [10]. The control strategy is developed in such a manner that while operating in grid-connected mode if there is any fault in the grid then there is a transition in the control strategy so that the reliability of supply is maintained by operating the inverters in an autonomous mode. The recent advancement in control strategies and the integration of information and communication technology (ICT) has facilitated the easy interconnection and management of these highly dispersed distributed energy resources. A brief description of GFM and GFL inverters is presented below.

The GFM inverters share the active and reactive power demand of the load among themselves in order to maintain the supply–demand balance. These inverters can be represented as an ideal AC voltage source with fixed frequency [11,12]. The inverters have hierarchical control architecture (primary, secondary, and tertiary layers) which will be explained in the later section. The primary control in GFM inverters is the droop control. Whereas, the distributed control strategy for the secondary control layer is very popular among these inverters as it enhances distributed intelligence and supports dynamic network topology. It also supports sparse communication hence more economic. Such control scheme does not suffer from single point failure in comparison to its traditional counterpart i.e., centralized control scheme. The control strategy is designed in such a manner to attain the objectives of frequency restoration, and proportional active (and/or) reactive power sharing. In addition, the voltage regulation can also be achieved by modifying the control scheme. Moreover, the distributed secondary control framework also supports plug and play capability.

The GFL inverters exchange the power produced, with the grid and offer various advanced grid support functions such as V–Q, f-P, fault ride-through, and many more. These inverters can be equivalently represented as an ideal current source connected in parallel to the AC grid. The GFL inverters are broadly classified as grid-feeding and grid-supporting inverters according to their modes of operation [13]. The grid-feeding inverters primarily exchange maximum active power with constant (typically zero) reactive power. However, the grid-supporting inverters provide ancillary services. During symmetrical and unsymmetrical voltage sags, the grid supporting inverters are designed to have the capability to detect such abnormalities and are mandated to be connected (known as low-voltage ride through in general) to support the grid. These inverters support the grid either through reactive power injection or negative-sequence compensation (during asymmetrical anomalies). The GFL inverters are integrated according to the IEEE Standard for Interconnection and Interoperability of Distributed Energy Resources (IEEE Std. 1547-2018) [14,15]. This standard mandates incorporation of at least one communication protocol from IEEE2030.5, IEEE 1815 (DNP3) and SunSpec Modbus. The communication interface has transformed the conventional grid into a smart grid facilitating bi-directional flow of information, distribution automation, energy management, demand-side response and maintenance activities like remote real-time health monitoring and fault diagnosis by manufacturers [16], which allows the ability to control and manage the available resources in a better way.

The work in this chapter is primarily related to investigate the working of the GFL inverters. Hence, there is a requirement to implement the corresponding control scheme on the testbed. There are several testbeds (like simulation, controller hardware-in-loop (CHIL), PHIL, power testbed, and full system) that are distinguished based on the variation in cost, fidelity, and coverage. The term '*cost*' is the total expense required to build deploy and maintain the testbed; the term '*fidelity*' means the proximity to the real-behavior by integration of various actual hardware devices, equipment and communication network and protocols to replicate the actual network latency and message packet behavior, etc.; and the term '*coverage*' defines the various tests which can be carried out on the developed testbed with safety. The comparison of these testbeds can be referred to [17]. In this work, we have presented real-time simulation testbed [17] and PHIL testbed [18] for the implementation of the control scheme for grid-connected three-phase inverters. The detailed system setup is described in this chapter with demonstration of some test cases.

The remainder of the chapter is summarized as: the different topologies of inverters are discussed in Section 5.2. It briefly describes each layer corresponding to the off-grid and grid-connected three-phase inverters. The hierarchical control structure is illustrated in Section 5.3. It also comprises of the detailed analysis of the control architecture of GFL inverters. The result and discussion along with the system under study are presented in Section 5.4. The development of real-time co-simulation testbed and PHIL testbed are also discussed in this section. Section 5.5 finally concludes the work.

5.2 Inverter topologies

As mentioned, the IIER can be operated broadly in two modes (i) grid-connected and (ii) islanded mode. They consist of both high-voltage (HV) and low-voltage (LV) parts. The control layer, device layer, and power devices are mutually integrated through the various communication interfaces. The device layer, on the one hand, models the control architecture to control the functionality of GFMs (or GFLs) and the control system, on the other hand, generates the power references required for power exchanges. A pictorial representation of different layers present in both these inverters is presented in the following subsections.

5.2.1 Grid forming inverters

Let us assume a GFM with a constant DC source (like PV, battery) as an input source. It consists of both high and low voltage parts as shown in Figure 5.1. The high-voltage part includes converters (like DC/DC and DC/AC) to convert one form of energy to another, PWM drivers to drive these converters and sensors (voltage and current) to measure the parameters for control actions. Whereas, the low-voltage part includes digital signal processors (DSPs) to implement the control

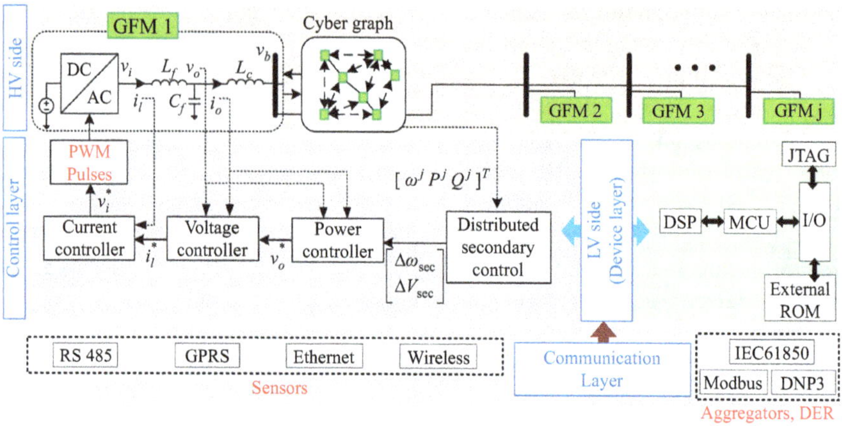

Figure 5.1 Layered architecture of an off-grid three-phase inverter system

schemes, micro-controller units (MCUs) to define the set-points fed to the DSPs, Joint Test Action Group (JTAG) to provide the access for testing/debugging the developed programs, external ROM for storing communication interface to interact between various devices. These inverters have the objectives of frequency restoration, and proportional active (and/or) reactive power sharing in a distributed control architecture. Each GFM receives the local as well as neighboring measurements to generate the control signals. The devices in various layers interact among themselves through the communication interface which can be either wired low-voltages. This control architecture also supports dynamic reconfiguration hence facilitating plug and play functionality. This feature is beneficial when certain resources are to be disconnected, or reconnected to the network.

It can be observed from Figure 5.1, that the network comprises of '*j*' number of GFMs in an autonomous mode. They communicate among themselves according to the cyber graph. The term '*cyber graph*' denotes the communication link present among the GFM inverters. These can be connected in any fashion like sparsely connected, partially-connected, or fully-connected. Moreover, these graphs can be dynamically configurated according to the system requirements. The represented system comprises primary droop control (along with the inner control loops for voltage and current) and distributed secondary control. The circuit elements and measurements are denoted by their usual representation and notations. The reference terms are denoted by superscript '*'. As the primary control scheme is itself not sufficient to maintain the zero steady-state error hence, the secondary controller is integrated into the system. The distributed secondary control, as the name suggests, obtains the frequency (ω), active (P), and reactive (Q) power measurements from its neighboring GFMs (according to the cyber graph). Afterwards, the correction terms corresponding to the voltage (ΔV_{sec}) and frequency ($\Delta\omega_{sec}$) are transmitted to the primary droop controller.

5.2.2 *Grid following inverters*

Apart from the components described for the GFM inverters, GFL inverters also consist of phase-locked loop (PLL) to synchronize with the grid. The layered-layout of GFL inverters is shown in Figure 5.2.

It can be observed from Figure 5.2, that the network that '*j*' number of GFLs are connected to the grid. The network elements and the measurements have the usual meaning. The designed control considered in the following system comprises of primary droop control scheme along with the inner control loop for current. The reference value is denoted by the superscript '*'. The PLL, on obtaining the voltage output, generates the frequency reference. The reference active (P) and reactive (Q) power are generated by the tertiary control, according to the load demand and available generation resources. Here, we have assumed constant real and reactive power reference values. Hence, tertiary control level modeling is neglected here. However, these can be included in advanced modeling designs to investigate the performance of various optimization schemes for economic dispatch, optimal power flow, unit commitment, etc. The grid support functions like V–Q, constant Q, P–Q for supporting voltages; constant P, f–P for restoring frequency; low voltage/frequency or high voltage/frequency ride through to support the grid during transients can further be incorporated. As mentioned, both GFM and GFL inverter architectures assume a constant DC source input. We can, however, include the intermittent behavior of renewable energy resources by integrating PV or wind sources. These can be integrated with additional control schemes like maximum power point tracking (MPPT), DC voltage link regulator, etc. These give a closer insight into the variation of available power from a renewable source and the DC voltage link fluctuation involved.

These inverters operate in (P/Q) mode i.e, exchanging active/reactive power with the grid according to the set points obtained from the control center in a centralized architecture. The system level architecture is shown in Figure 5.3.

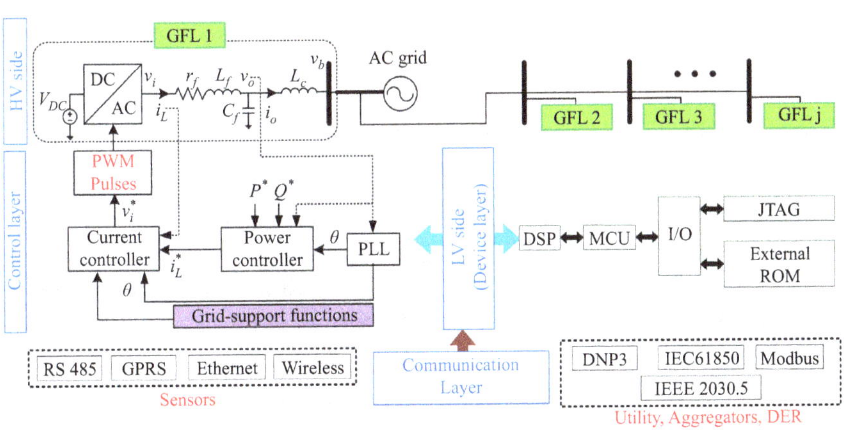

Figure 5.2 Layered architecture of a grid-tied three phase inverter system

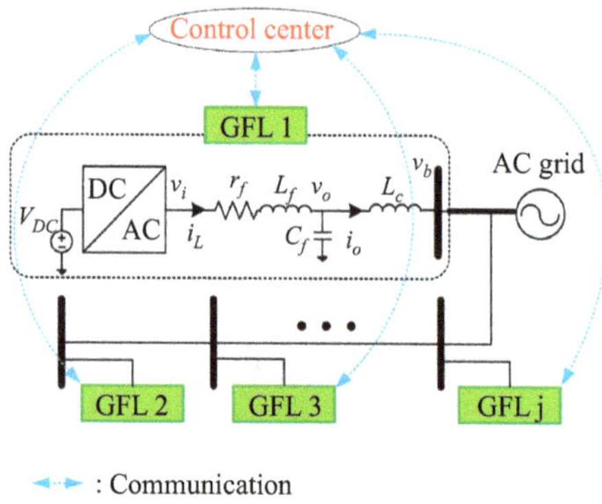

Figure 5.3 Centralized P/Q control mode

Here, each GFL is sharing its measurements with the centralized control center and receives the control commands to exchange active and (or) reactive power references according to the load demand in the network. The interaction between these devices at various levels is carried out through the communication interface (like wireless, Ethernet-based, GPRS, RS485, etc.). Apart from the internal control functions of converters within the device layer through SunSpec Modbus and IEC 61850 protocol; these interfaces also facilitate the grid-support functionalities at the system level through IEEE Std. 2030.5 standard, DNP3 as shown in Figure 5.2. At the system level, the interconnection between the utility, DER clients, inverter manufacturers, and aggregators is beneficial to carry out energy management schemes, real-time monitoring etc. Additionally, the performance evaluation and fault diagnosis can also be executed for the devices through these interfaces.

The grid-operators regulate the voltage and frequency of the grid according to the standards and recommendations [14,15]. The various functions and the associated modes can be summarized as:

- The grid-voltage is supported by adjusting the real/reactive power exchanges. The different modes supported are: constant power factor (PF) mode, V–Q mode, constant Q mode, P–Q mode, V–P mode as shown in Figure 5.4.
- The grid-frequency is restored by regulating active power exchanges. The various modes are: maximum P mode, constant P mode, and f–P mode as shown in Figure 5.4.
- During transient disturbances, the low/high voltage/frequency ride through functionalities are provided to support the grid voltage and frequency as shown in Figure 5.4. These are distinctly defined for various countries according to the allowable voltage and frequency limits.

As mentioned, the large integration of these renewable energy resources into the grid may cause frequency instability and reliability issues. Hence, these grid support functions are beneficial to combat the transient instability which may arise due to upcoming disturbances.

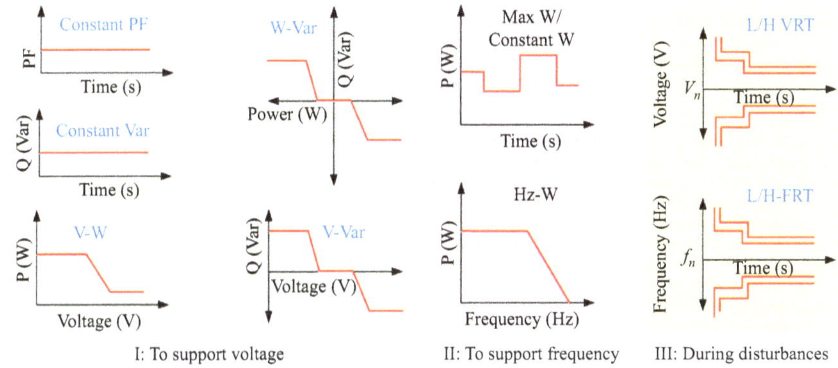

Figure 5.4 Various grid support functions and the associated modes

5.3 Control strategies

There are various intelligently designed control strategies for different configurations of inverters. For GFL inverters, although the voltage and frequency are maintained by the grid, a control framework is required for real and reactive power exchange [19]. The control operation can also manage the harmonics to be injected and the power quality to be maintained of the exchanged power. On the contrary, in GFM inverters, the control framework is required to regulate all the parameters such as voltage, frequency, active, and reactive power in the system [20]. The hierarchical control architecture of the IIERs includes three layers namely primary, secondary and tertiary control layers. The features of each layer like objectives, control methods, algorithms and time scale of operation are shown in Table. 5.1 [21–23]. Depending on the type of inverter connected to the network, the control architecture can be centralized, decentralized, or distributed. For instance, the centralized and decentralized control framework is popular among the grid-connected IIERs and distributed control framework is highly adopted for autonomous systems connected with IIERs but can also be considered for GFLs [24].

It can be observed that the primary control layer has the smallest time of operation, hence the fastest one. Whereas the tertiary control layer is the slowest among all the layers. Hence, the functions of these layers are also allotted accordingly. The GFM inverters comprise distributed secondary control architecture whose detailed mathematical formulation and implementation can be obtained from [17,25]. The detailed illustration of the control architecture of GFL inverters is explained in the following subsection.

5.3.1 Control architecture of GFL inverters

Let us consider a constant DC source at the input of grid-tied three-phase inverter which is connected to the grid at point of common coupling (PCC) as shown in Figure 5.5. Therefore, this section will discuss the control architecture of the DC/AC

Table. 5.1 Features of hierarchical control layer

Layer	Features
Primary layer	(i) It is responsible for converter output control and active (and/or) reactive power sharing. (ii) The control schemes like droop control strategy or communication-based strategy are utilized. (iii) For converter output control algorithms can be implemented through synchronous frame, stationary frame, or natural frame. For power sharing control, the algorithms included in the droop based control are conventional droop (P–f/Q–V), modified droop (V–P/f–Q), hybrid droop, virtual impedance, enhanced virtual output impedance loop-based, virtual-inertia-based, angle-droop, Q–V derivative droop and adaptive droop. Whereas, power sharing with the integration of communication is master–slave, active load sharing, or circular chain. (iv) The time-scale of operation is within tens of milliseconds and has the fasted response time.
Secondary layer	(i) It comprises of the objectives of frequency restoration, proportional active (and/or) reactive power sharing, and voltage regulation. Recently the economic dispatch is also included in this layer for faster response. It is used for real-time load management, secondary load-frequency control and automatic generation control. (ii) The centralized and decentralized control schemes are adopted. However, distributed control strategy is popular for GFM inverters. (iii) The centralized control scheme includes algorithms such as PI control-based, potential function-based, model-predictive control (MPC)-based, hybrid MPC based, and compensation-based. The different control algorithms for distributed control schemes are multi-agent-based, consensus-based, distributed MPC-based, distributed cooperation-based, heuristic approach-based, decomposition-based and hybrid control. (iv) The time-scale of operation is within hundreds of milliseconds, hence has the response time in between the other two control layers.
Tertiary layer	(i) It is used for voltage Var control, unit commitment, optimal power flow and economic dispatch. (ii) The centralized and decentralized control schemes are adopted. However, the centralized control strategy is popular among GFL inverters. (iii) Heuristic optimization algorithms like genetic algorithm (GA) and particle swarm optimization (PSO), model-based like MPC etc. (iv) The time-scale of operation is within seconds to minutes and has the slowest control operation.

converter to exchange the reference active/reactive power with the grid [26]. The overall control framework is in d–q frame because of its simplicity of implementation and the efficiency of operation. The main elements of the control architecture are (i) PLL, (ii) power controller and (iii) current controller as depicted in Figure 5.5 [27]. The associated models of each element are briefly described in the subsequent subsections.

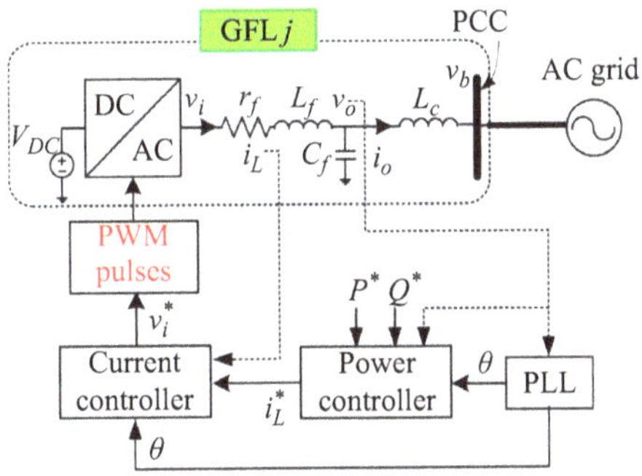

Figure 5.5 The block representation of the control scheme

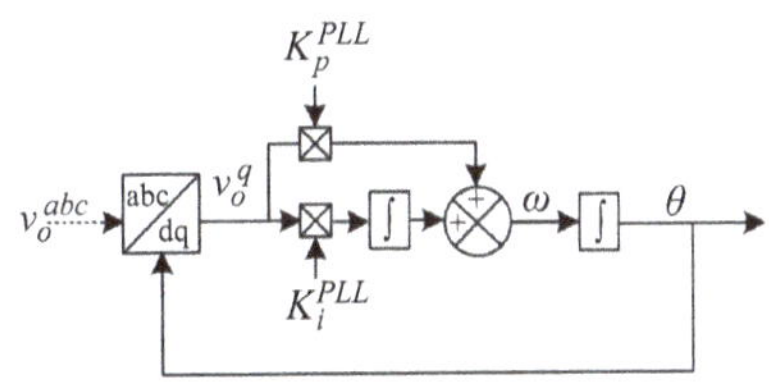

Figure 5.6 Model of PLL

5.3.2 PLL

In order to synchronize the grid-tied inverter to the grid, PLL is mandatory. The model of PLL is represented in Figure 5.6. The PLL logic considered here assumes the aligning of closed-loop control angle of the dq-transformation in such a manner that there is no q-axis component at PCC.

5.3.3 Power controller

Its objective is to follow the real/reactive power set points (P^*/Q^*) to regulate the active/reactive power exchanges with the grid. These power references and the voltage at PCC are used to generate references for the current controller. Hence, it can be precisely expressed as an open-loop converter to transform the power references to the current references as shown in Figure 5.7.

It is shown in Figure 5.5 that the output of the inverter is connected through the LC filter with an additional inductor to the PCC. The current which needs to be controlled is the inductor current to follow the generated current references (or in turn the power references). Therefore, the reference current can be represented as:

$$\sum i^d = i_o^{d^*} + i_C^d = i_o^{d^*} + \left(i_L^d - i_o^d\right) \tag{5.1}$$

$$\sum i^q = i_o^{q^*} + i_C^q = i_o^{q^*} + \left(i_L^q - i_o^q\right) \tag{5.2}$$

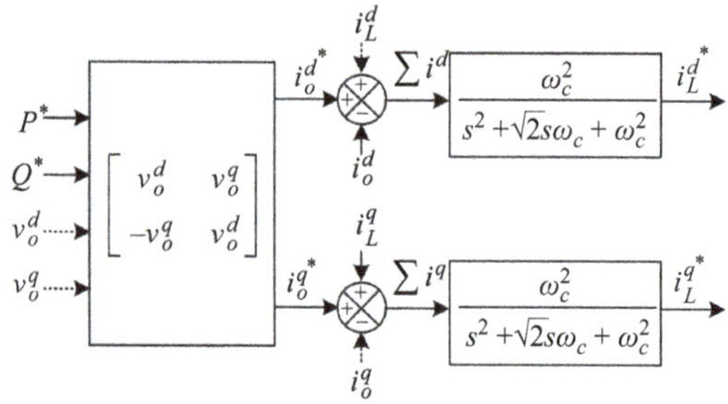

Figure 5.7 Model of power controller

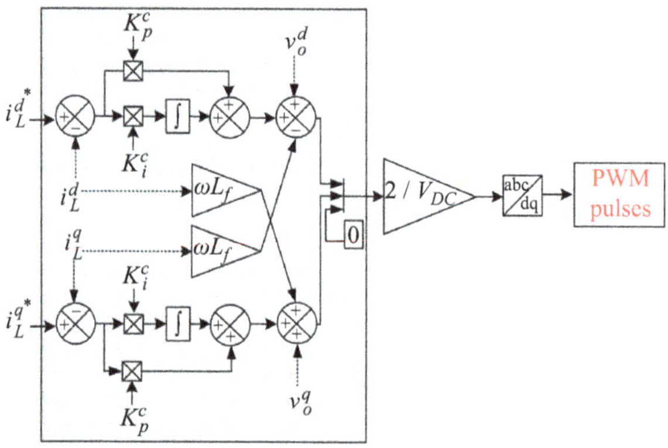

Figure 5.8 Model of current controller

Moreover, to remove the harmonics and noise which might have appeared in these current references due to the distorted voltage at the PCC, these references are passed through low pass filter (generally represented as a second-order Butterworth low-pass transfer function) [26] as shown in Figure 5.7.

5.3.4 Current controller

The current controller is represented as a pair of PI controllers as shown in Figure 5.8 [28]. It has both decoupling and feed-forward terms for PCC voltage. As the gain (on average) of the inverter is equivalent to the DC link voltage, so to normalize it, the division with DC link voltage is included. This takes into account the variations in DC link voltage.

5.4 Results and discussion

An inverter with the constant DC source interconnected with the grid at 400 V (l-l rms) and the nominal frequency of 50 Hz (as shown in Figure 5.5) is considered as a test system under study. The specifications of the inverter, network, load and control parameters are tabulated in Table 5.2.

The implementation of the considered grid-tied three-phase inverter has been carried out in both the real-time cosimulation testbed and the PHIL testbed. The components involved in both these testbeds are illustrated in the subsequent subsections.

5.4.1 Real-time co-simulation testbed

The real-time co-simulation testbed [17] as shown in Figure 5.9 comprises of OP-5700 RT simulator to model the cyber-physical framework of the grid-tied three-phase inverter. The physical layer with the involved control schemes is implemented in the HYPERSIM software on the test personal computer (PC). The SEL-3530 real-time automation controller (RTAC) hardware integrated with ACCELERATOR RTAC

Table 5.2 System and control parameters

Parameter	Symbol	Value
DC link voltage	V_{DC}	750 V
LC filter	L_f; r_f; C_f	1.35 mH; 0.05 Ω; 50 μF
Coupling impedance	L_c	0.3 mH proportional gain of PLL K_p^{PLL} 2.1
Integral gain of PLL	K_i^{PLL}	5,000
Proportional gain of current controller	K_p^c	1
Integral gain of current controller	K_i^c	460
Switching frequency	f_s	8 kHz
Active power reference	P^*	10 kW
Reactive power reference	Q^*	5 kVar

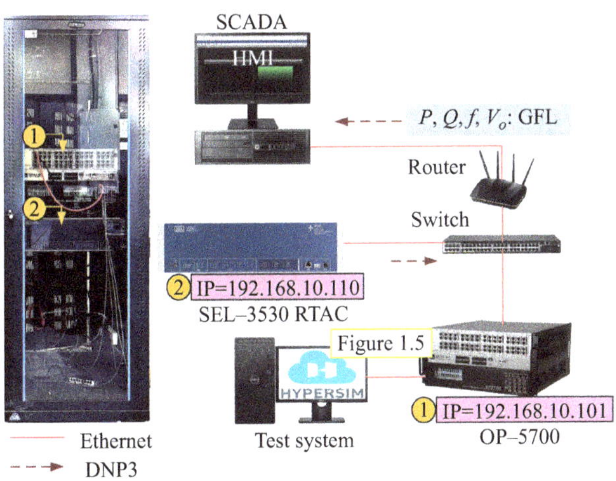

Figure 5.9 Real-time co-simulation testbed

SEL-5033 software is used to implement the cyber layer for monitoring purposes. It is used for the development of human–machine interface (HMI) through ACCELERATOR Diagram Builder SEL-5035 software for local/web-based remote monitoring and control. The distributed network protocol 3 (DNP3) is used for HMI and interconnects the physical and cyber layers to monitor the parameters (like frequency, power, voltage, etc.) of the system. The OP-5700 RT simulator, RTAC hardware, and test PC are connected through switch on Ethernet-based interface.

The results captured for frequency, voltage, real, and reactive power on the real-time co-simulation testbed are shown in Figure 5.10. Moreover, the responses to the change in power references are also plotted. These verify the effectiveness of the developed control scheme which tracks the reference real/reactive power. The voltage and the frequency are also following their nominal values.

5.4.2 Power hardware-in-loop testbed

This testbed comprises of a real inverter, typhoon HIL simulator, RTDS simulator, real resistive load, and four-quadrant power amplifier as shown in Figure 5.11. The

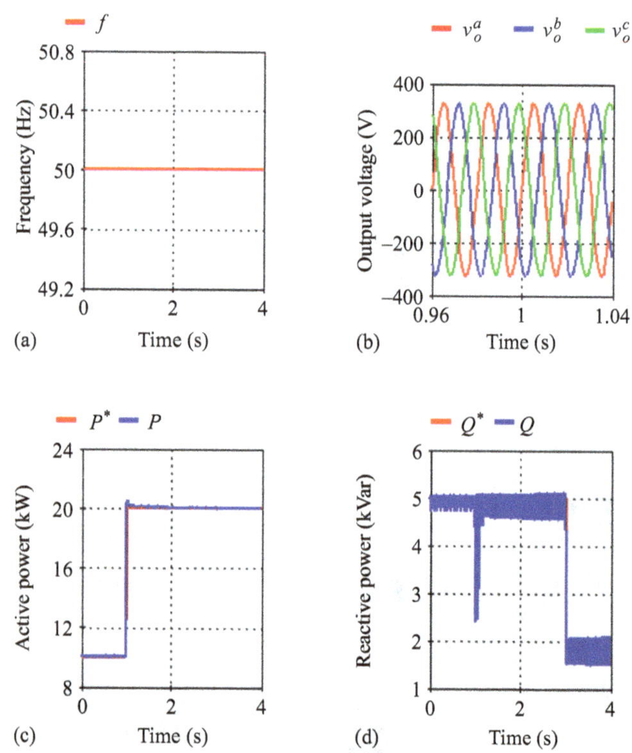

Figure 5.10 Real-time simulation results on HYPERSIM with variation of active power reference at 1 s and reactive power reference at 3 s for (a) frequency; (b) voltage (zoomed); (c) real power; and (d) reactive power

control scheme of this real inverter is deployed in the typhoon HIL simulator, whereas the three-phase grid is simulated in the RTDS simulator. The four-quadrant amplifier is an intermediate device to connect the devices at power and signal level. The overall arrangement of these devices for the implementation of the control scheme is presented in Figure 5.12.

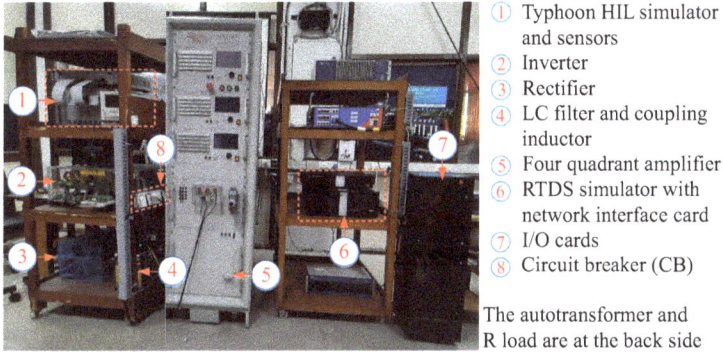

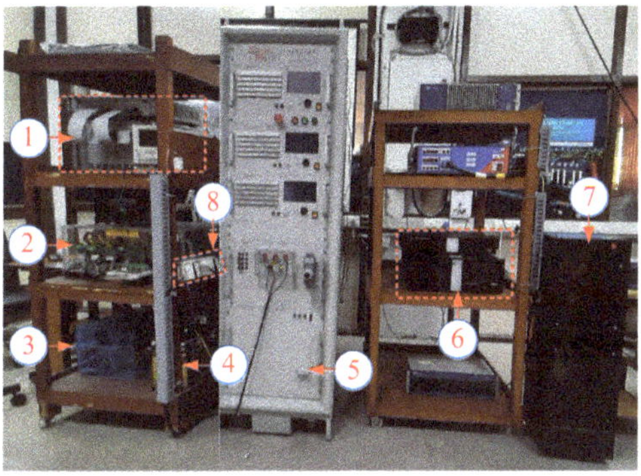

Figure 5.11 The PHIL testbed

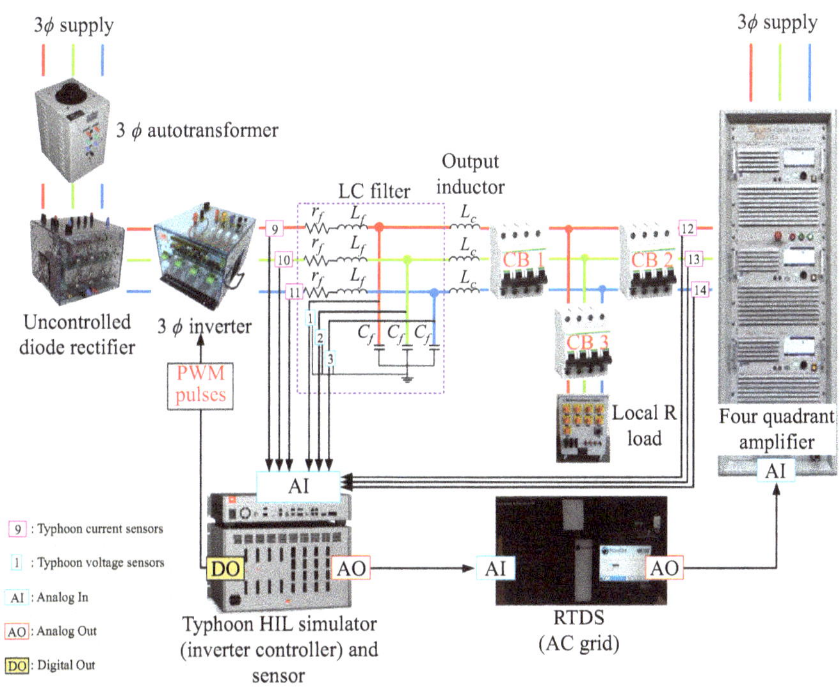

Figure 5.12 The overall connection diagram of all components in the PHIL testbed

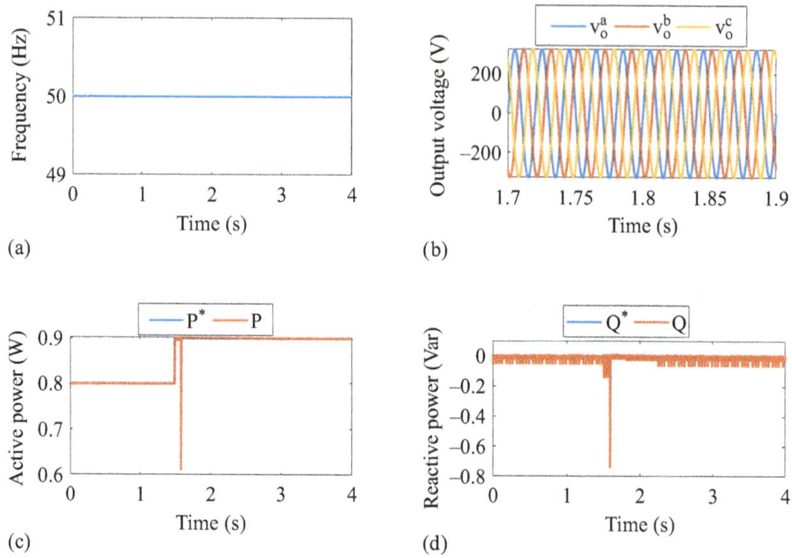

Figure 5.13 The captured signals on the PHIL testbed with variation of active power reference at 1 s for (a) frequency; (b) voltage (zoomed); (c) real power; and (d) reactive power

The captured graphs for frequency, voltage, real, and reactive power are plotted on the PHIL testbed as shown in Figure 5.13. The active power references are varied at 1 s. The plots certify the efficacy of the implemented control scheme of the grid-tied three-phase inverter.

5.5 Conclusion

The IIERS can be operated in grid-connected or autonomous mode. A brief overview of the functionality and control of these inverters is presented. However, the work in this chapter is dedicated to presenting a detailed description of the grid-tied three-phase inverters. These inverters are used to exchange power (active and/or reactive) and also participate in the ancillary services. The associated control scheme is presented in detail and is further implemented on two testbeds i.e., real-time co-simulation testbed and PHIL testbed. The distinct features of these testbed are summarized in the chapter. Further, the results on the developed testbeds verify the effectiveness of the control scheme. The active (and/or) reactive power references are precisely captured by the presented control framework. As mentioned, similar to the control architecture of grid-tied three-phase inverters, several other case studies with different system models, control and communication architectures can be investigated on these testbeds. Such implementations help the researchers to gain better insights into the system.

References

[1] B. Mirafzal and A. Adib, "On grid-interactive smart inverters: features and advancements," *IEEE Access*, vol. 8, pp. 160526–160536, 2020, doi:10.1109/ACCESS.2020.3020965.

[2] G. Simard, *IEEE Grid Vision 2050*, pp. 1–93, 2013, doi:10.1109/IEEESTD.2013.6577603.

[3] S. Gangavarapu, M. Verma and A. K. Rathore, "A novel transformerless single-stage grid-connected solar inverter," *IEEE J. Emerg. Sel. Topics Power Electron.*, vol. 11, pp. 970–980, 2023, doi:10.1109/JESTPE.2020.3007556.

[4] D. Ramasubramanian, P. Pourbeik, E. Farantatos, and A. Gaikwad, "Simulation of 100% inverter-based resource grids with positive sequence modeling," *IEEE Electrific. Mag.*, vol. 9, no. 2, pp. 62–71, 2021, doi:10.1109/MELE.2021.3070938.

[5] Y. Cheng, L. Fan, J. Rose, *et al.*, "Real-world subsynchronous oscillation events in power grids with high penetrations of inverter-based resources," *IEEE Trans. Power Syst.*, vol. 38, pp. 316–330, 2022, doi:10.1109/TPWRS.2022.3161418.

[6] A. Hoke, V. Gevorgian, S. Shah, P. Koralewicz, R. W. Kenyon and B. Kroposki, "Island power systems with high levels of inverter-based resources: stability and reliability challenges," *IEEE Electrific. Mag.*, vol. 9, no. 1, pp. 74–91, 2021, doi:10.1109/MELE.2020.3047169.

[7] Y. Li, Y. Gu, and T. Green, "Revisiting grid-forming and grid-following inverters: a duality theory," *IEEE Trans. Power Syst.*, vol. 37, pp. 4541–4554, 2022, doi:10.1109/TPWRS.2022.3151851.

[8] W. Du, F. Tuffner, K. P. Schneider, *et al.*, "Modeling of grid-forming and grid-following inverters for dynamic simulation of large-scale distribution systems," *IEEE Trans. Power Del.*, vol. 36, no. 4, pp. 2035–2045, 2021, doi:10.1109/TPWRD.2020.3018647.

[9] B. Pawar, E. I. Batzelis, S. Chakrabarti, and B. C. Pal, "Grid-forming control for solar PV systems with power reserves," *IEEE Trans. Sustain. Energy*, vol. 12, no. 4, pp. 1947–1959, 2021, doi:10.1109/TSTE.2021.3074066.

[10] X. Hou, Y. Sun, J. Lu, *et al.*, "Distributed hierarchical control of AC microgrid operating in grid-connected, islanded and their transition modes," *IEEE Access*, vol. 6, pp. 77388–77401, 2018, doi:10.1109/ACCESS.2018.2882678.

[11] S. Anttila, J.S. Döhler, J. G. Oliveira, and C. Boström. "Grid forming inverters: a review of the state of the art of key elements for microgrid operation," *Energies*, vol. 15, no. 15, p. 5517, 2022.

[12] R. H. Lasseter, Z. Chen, and D. Pattabiraman, "Grid-forming inverters: a critical asset for the power arid," *IEEE J. Emerg. Sel. Topics Power Electron.*, vol. 8, no. 2, pp. 925–935, 2020, doi:10.1109/JESTPE.2019.2959271.

[13] B. Mirafzal and A. Adib, "On grid-interactive smart inverters: features and advancements," *IEEE Access*, vol. 8, pp. 160526–160536, 2020, doi:10.1109/ACCESS.2020.3020965.

[14] California Public Utilities Commission, "Recommendations for updating the technical requirements for inverters in distributed energy resources, smart inverter working group recommendations," 2014.

[15] IEEE Standard for Interconnection and Interoperability of Distributed Energy Resources With Associated Electric Power Systems Interfaces, Standard 1547–2018, Apr. 2018.

[16] F. R. Yu, P. Zhang, W. Xiao, and P. Choudhury, "Communication systems for grid integration of renewable energy resources," *IEEE Network*, vol. 25, no. 5, pp. 22–29, 2011.

[17] K. Gupta, S. Sahoo, B. K. Panigrahi, F. Blaabjerg, and P. Popovski, "On the assessment of cyber risks and attack surfaces in a real-time co-simulation cybersecurity testbed for inverter-based microgrids," *Energies*, vol. 14, no. 16, p. 4941, 2021, doi:10.3390/en14164941.

[18] P. C. Kotsampopoulos, V. A. Kleftakis and N. D. Hatziargyriou, "Laboratory education of modern power systems using PHIL simulation," *IEEE Trans. Power Sys.*, vol. 32, no. 5, pp. 3992–4001, 2017, doi:10.1109/TPWRS.2016.2633201.

[19] D. B. Rathnayake and B. Bahrani, "Multivariable control design for grid-forming inverters with decoupled active and reactive power loops," *IEEE Trans. Power Electron.*, vol. 38, pp. 1635–1649, 2022, doi:10.1109/TPEL.2022.3213692.

[20] D. B. Rathnayake, C. Phurailatpam, S. P. Me, *et al.*, "Grid forming inverter modeling, control, and applications," *IEEE Access*, vol. 9, pp. 114781–114807, 2021, doi:10.1109/ACCESS.2021.3104617.

[21] D. E. Olivares, A. Mehrizi-Sani, A. Etemadi, *et al.*, "Trends in microgrid control," *IEEE Trans. Smart Grid*, vol. 5, no. 4, pp. 1905–1919, 2014, doi:10.1109/TSG.2013.2295514.

[22] Y. Yao and N. Ertugrul, "An overview of hierarchical control strategies for microgrids," *2019 29th Australasian Universities Power Engineering Conference (AUPEC)*, pp. 1–6, doi:10.1109/AUPEC48547.2019.211804.

[23] S. Shrivastava and B. Subudhi, "Comprehensive review on hierarchical control of cyber-physical microgrid system", *IET Gener. Transm. Distrib.*, vol. 14, no. 26, pp. 6397–6416, 2020.

[24] A. Singhal, T. L. Vu, and W. Du, "Consensus control for coordinating grid-forming and grid-following inverters in microgrids," *IEEE Trans. Smart Grid*, vol. 13, no. 5, pp. 4123–4133, 2022, doi:10.1109/TSG.2022.3158254.

[25] K. Gupta, S. Sahoo, R. Mohanty, B. K. Panigrahi, and F. Blaabjerg, "Decentralized anomaly characterization certificates in cyber-physical power electronics based power systems," in: *IEEE 22nd Workshop on Control and Modelling of Power Electronics (COMPEL)*, 2021, pp. 1–6, doi:10.1109/COMPEL52922.2021.9645984.

[26] N. Kroutikova, C. A. Hernandez-Aramburo, and T. C. Green, "State-space model of grid-connected inverters under current control mode," *IET Elect. Power Appl.*, vol. 1, no. 3, pp. 329–338, 2007.

[27] A. Yazdani and R. Iravani, *Voltage-Sourced Converter in Power Systems: Modelling, Control, and Application.* New York, NY: Wiley, 2010.

[28] N. Pogaku, M. Prodanovic, and T. C. Green, "Modeling, analysis and testing of autonomous operation of an inverter-based microgrid," *IEEE Trans. Power Electron.*, vol. 22, no. 2, pp. 613–625, 2007, doi:10.1109/TPEL.2006.890003.

Chapter 6
Sliding mode control of a three-phase inverter

Youssef Errami[1], Abdellatif Obbadi[1], Smail Sahnoun[1] and Mohammed Ouassaid[2]

This chapter proposes a sliding mode approach (SMA) for voltage source inverter (VSI) to regulate the powers injected into the grid. A VSI is employed to connect the wind power system (WPS) to the electrical network system (ENS) and to process the energy generated by this system. The SMA is used for both the three-phase inverter and the rectifier. The inverter is commanded to control the delivered power to the ENS and to sustain invariable the voltage of the DC-link, whereas the rectifier is controlled to guarantee the maximum power point (MPP) for the wind turbine (WT). So, the active current reference is produced via an external loop that has the function of keeping constant voltage of DC-link, but the reactive current reference is fixed to zero to guarantee unity power factor (UPF). The stability of the regulators is achieved via Lyapunov analysis. Simulation results confirm the success of the presented approach and show it can work consistently under diverse conditions. Also, it reduces the impact of short-circuit and fluctuations of voltage on the grid. The comparative results and analysis for the presented SMA strategy and the conventional vector control (VC) are presented for different grid voltage conditions to indicate the excellent performances of SMA method including smoother transient responses and less fluctuation under fault conditions.

6.1 Introduction

Three-phase VSI is the most interesting dc–ac system of several complex applications for instance smart grid (SG), renewable energy systems (RESs), uninterruptable power equipment, flexible AC transmission systems, and active power filter (APF) [1,2]. On the other hand, the performance of VSI is principally determined by command methods. So, the control of VSI has been largely studied and recent techniques are proposed every year. Numerous controllers have been

[1]Laboratory of Electronics, Instrumentation and Energy, Department of Physics, Faculty of Sciences, University Chouaib Doukkali, Morocco
[2]Engineering for Smart and Sustainable Systems Research Center, Mohammadia School of Engineers, Mohammed V University in Rabat, Morocco

presented in the literature to control inverters. vector control technique (VCT) based on proportional integral (PI) regulators were employed for VSI. Hamid *et al.* [3] and Shuhui *et al.* [3,4] have proposed MPP from the RESs-based synchronous generators under deviation of wind velocity. Salvador *et al.* [5] have presented a fault ride-through (FRT) approach of the RESs all through the ENS fault and if the voltage of the network was totally interrupted. The conventional vector control (VC) decouples the current of the converter into active and reactive power components. So, network current components are used to realize reactive and active power control. VC is simple to apply but it is characterized by its reduced aptitude to accept the results of the parameter changes which can degrade the control qualities. Moreover, despite the considerable gain correction, it can degrade the transitory performance owing to the trade-off between preserving the stability of the system over the entire functioning range and attaining a satisfactory dynamic response through transient responses [2]. An alternative to the VC approach is the direct power control (DPC) method. So, it can be used to control the inverter and to overcome the drawbacks of VC strategy. Indeed, the core idea of this technique is the use of the look-up tables and hysteresis comparators. Also, it is based on reducing the differences between the reference and real value of reactive/active power [6]. In [7], Noguchi *et al.* proposed a method based on pulse width modulation (PWM) to enhance the total power factor and effectiveness. The powers can be commanded directly with numerous comparators and a switching table. Also, the evaluation technique of the voltage is founded on an estimation of the instantaneous powers for each switching situation of the system. Malinowski *et al.* [8] presented a direct power command of converters via constant switching frequency and SVM. For this purpose, the virtual flux estimator substituted line voltage sensors in this work. As a result, it exhibits numerous qualities, such as an easy technique, excellent dynamic response, and constant switching frequency. Therefore, DPC is accomplished without an internal current loop. Besides, DPC has a very quick dynamic response. But, DPC creates large steady power undulations and changeable switching frequency as a result of the employ of hysteresis comparators and look-up tables. Several works have reported the employ of other techniques to control power converters. In [9], Evran proposed an inverter topology with two stage for grid-connected power system by operating a converter to connect the system with low-voltage to the electrical network and to diminish the THD. Xiao *et al.* [10] used the controllers for dc-bus voltage via diverse references for an electrical network PV converter functioning in low and normal grid voltage modes. Also, a supplementary DC-bus voltage and MPP regulator are employed to ensure the efficiency of the low voltage ride through (LVRT) of the power electronic converter. Rachid *et al.* [11] presented the design of a disturbance observer via a control for inverters with a boost converter taking into account unbalanced grid voltages. In addition, they used diverse controllers with a disturbance observer approach technique with a feedback linearizing approach to control the powers fed into the power network.

Recently, the SMA has been widely used because of its significant advantages such as the fast dynamic response, assured stability, disturbance rejection ability, and strong robustness against uncertainties, parameter variations, and errors of modeling. In [12], a switching control law is employed in the SMA to apply to unipolar single-phase inverters and to drive the uncertainties of the VSI onto the sliding surface in order to sustain its path on the sliding surface for the following time. Ramon *et al.* [13] proposed an SMA based on a Kalman filter (KF) to control LCL-filtered three-phase inverter connected to the grid. Then, this approach permits to give the required dynamics to the grid-side current for direct control of the current fed into the grid without the grid harmonics compensation. Besides, the proposed method develops the tracking quality of the reference and enhances the robustness rewarding variation of network inductance. Merabet *et al.* [14] used a robust feedback linearizing approach-based SMA in the converter to improve its robustness to uncertainties under grid faults. So, this technique was developed from the grid and the DC-link models. Yunkai *et al.* [15] have used a disturbance observer using SM and fuzzy logic techniques to command grid connected converter and to estimate, in real-time, the disturbances of the system. So, the converter has strong robustness as the perturbations can be recompensed in an adaptive manner. Said *et al.* [16] proposed robust adaptive SMA for a boost converter using an indefinite external input voltage and resistive load. So, the authors employed state estimators and promise the stability of the closed-loop systems. Also, we can observe the convergence of the estimated values of the load resistances and the input voltage toward their true values. Rodrigo *et al.* [17] presented a multi-loop technique SMC to guarantee the tracking of the current for VSI. For this purpose, a first loop is implemented in order to track a particular current reference and a second closed loop acting as a current source is used to drive a capacitive+ inductive (CL) circuit without depending on the grid side voltage and impedance. In [18], the authors presented a robust control method for wind RESs using an inverter exclusive of a transformer and which is coupled to the network via an LCL filter. So, two SM regulators are used to control the powers. In addition, a PI controller is utilized, in the external loop of voltage, to control the voltage of the DC-link and produce the necessary currents for the internal current-loop in the inverter. Tingting *et al.* [19] proposed a three-phase converter to attain dynamic performances and improved stability. Then, the control scheme and the regulators are designed using the model of the system and this method can decrease the effects of the variations of the resistance load and the output voltage demand.

In this context, this chapter presents SMA for VSI to regulate the powers injected in the network. A VSI is utilized to connect the WPS-based permanent magnet synchronous generator (PMSG) to the electrical grid and to process the energy generated by this system. For this purpose, VSI control can ensure the stability of DC bus voltage and the synchronization of the power, delivered by the system, with the grid. Besides, it should have the ability to adjust the powers exchanged with the power network, to guarantee extracting desired power from the wind system, transfer high quality current to the electrical network to conform to grid interconnection standards and guarantee the maximum possible power factor of dispositive. The stability of the controllers is attained with Lyapunov method. Simulation results confirm the efficacy of the presented approach and demonstrate it can work efficiently in diverse conditions.

6.2 Modeling description and control of the inverter

6.2.1 Mathematical model of the DC/AC converter

Figure 6.1 presents the configuration of a VSI with an RL filter composed of six power switching devices, supplied by a constant dc power supply and an output RL filter (Rf and Lf). The presented SMA of the converter is obtained by using the transformation of Park. The equations can be obtained from Figure 6.1 as:

$$\begin{bmatrix} e_a \\ e_b \\ e_c \end{bmatrix} = R_f \begin{bmatrix} i_a \\ i_b \\ i_c \end{bmatrix} + L_f \frac{d}{dt} \begin{bmatrix} i_a \\ i_b \\ i_c \end{bmatrix} + \begin{bmatrix} v_a \\ v_b \\ v_c \end{bmatrix} \tag{6.1}$$

e_a, e_b, e_c are voltages at the output of VSI; v_a, v_b, v_c are voltage components of grid; i_a, i_b, i_c are line currents; L_f is the filter inductance; R_f is the filter resistance.

Using the transformation of Park in (6.1) yields:

$$\frac{di_{d-f}}{dt} = -\frac{R_f}{L_f} i_{d-f} + \omega i_{q-f} + \frac{1}{L_f} e_d - \frac{1}{L_f} v_d \tag{6.2}$$

$$\frac{di_{q-f}}{dt} = -\frac{R_f}{L_f} i_{q-f} - \omega i_{d-f} + \frac{1}{L_f} e_q - \frac{1}{L_f} v_q \tag{6.3}$$

where

e_d, e_q is the output voltages of the VSI; v_d, v_q is the output voltages of the grid; i_{d-f}, i_{q-f} are the components of grid currents; ω is the grid frequency.

The grid frequency is determined by PLL. At the DC-link :

$$C\frac{dU_{dc}}{dt} = i_0 - i_{dc} \tag{6.4}$$

U_{dc} is the voltage of DC-link; i_{dc} is current at the input of the inverter; i_0 is the current at the output of the generator; C is the DC-link capacitor.

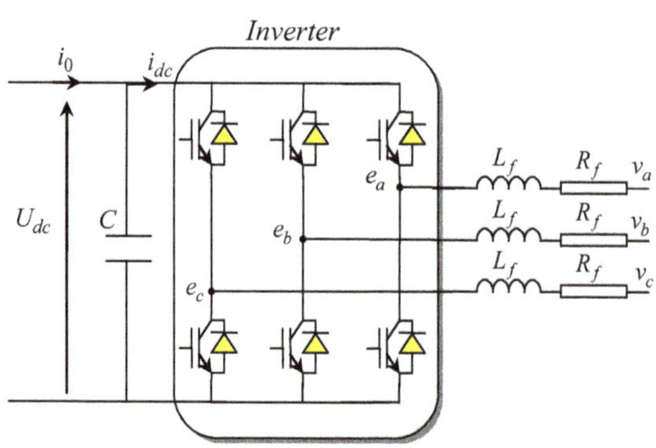

Figure 6.1 Schematic diagram of VSI

Also, the power is:

$$U_{dc} i_0 = \frac{3}{2}\left(v_d i_{d-f} + v_q i_{q-f}\right) \tag{6.5}$$

From (6.4) and (6.5), yields:

$$\frac{dU_{dc}}{dt} = \frac{3}{2C}\left(\frac{v_q}{U_{dc}} i_{q-f} + \frac{v_d}{U_{dc}} i_{d-f}\right) - \frac{1}{C} i_{dc} \tag{6.6}$$

Then, we choose the following axis orientation:

$$v_d = U_0 \text{ and } v_q = 0 \tag{6.7}$$

The dynamics (6.6) can be rewritten as

$$\frac{dU_{dc}}{dt} = \frac{3}{2C}\frac{U_0}{U_{dc}} i_{d-f} - \frac{1}{C} i_{dc} \tag{6.8}$$

Also, we can rewrite (6.2) and (6.3) as:

$$\frac{di_{d-f}}{dt} = -\frac{1}{L_f} U_0 - \frac{R_f}{L_f} i_{d-f} + \omega i_{q-f} + \frac{1}{L_f} e_d \tag{6.9}$$

$$\frac{di_{q-f}}{dt} = -\frac{R_f}{L_f} i_{q-f} - \omega i_{d-f} + \frac{1}{L_f} e_q \tag{6.10}$$

Furthermore, the powers can be written as:

$$P_{grid} = \frac{3}{2} U_0 i_{d-f} \tag{6.11}$$

$$Q_{grid} = \frac{3}{2} U_0 i_{q-f} \tag{6.12}$$

From (6.11) and (6.12), it is possible to prove that by controlling the injected grid-currents, i_{d-f} and i_{q-f}, the powers injected into the grid can be handled. Besides, the dynamic references design will permit to define appropriate power references to find the equivalent grid current reference using (6.11) and (6.12).

6.2.2 Proposed SMA

Usually, SMA can be developed in two stages phases. The first phase is to select the surface of the SMA. The next phase consists in establishing the input of the control that will guarantee the forcing of the system trajectory towards the SMA surface, which guarantees the system to assure the following SMA accomplishment condition:

$$S\frac{dS}{dt} \prec 0 \tag{6.13}$$

S is the SMA surface.

If we assume that the voltage of the DC link is previously stable and effectively controlled ($U_{dc} = U_{dc\text{-}ref}$), the relationship between i_{d-f} and i_0 can be provided by

$$U_{dc\text{-}ref} i_0 = \frac{3}{2} U_0 i_{d-f} \tag{6.14}$$

Consequently, the dynamic model (8) can be expressed as:

$$\frac{dU_{dc}}{dt} = \frac{3}{2C} \frac{U_0}{U_{dc\text{-}ref}} i_{d-f} - \frac{1}{C} i_{dc} \tag{6.15}$$

Equation (6.15) involves that the command of the DC-link voltage can be accomplished by commanding i_{d-f}. On the other hand, (6.11) and (6.12) present the laws of the current internal loop, which is established on the fact that the powers can be commanded by the dq-axis currents. The DC/AC control is realized via $d-q$ components of network current, as it is depicted in Figure 6.2. The reference of the frequency/phase for performing Park transformation is given using a PLL. The control strategy uses two closed-loop commands to control the reactive power and the DC link voltage separately. The DC-link voltage reference $U_{dc\text{-}ref}$ is the input of the external loop controller to control the converter output active power, while the output is the d-axis reference current i_{dr-f}. The outer-loop voltage controller provides a reference for the inner-loop current controller based SMA and which is the key to control active and reactive powers. Indeed, slow dynamic is associated with the proportional integral (PI) regulator to produce the reference source current

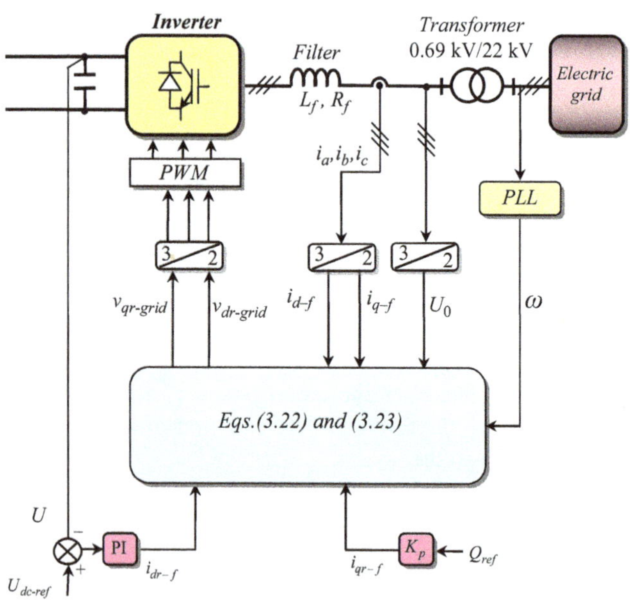

Figure 6.2 Block diagram of the proposed SMC for inverter

i_{dr-f}. In addition, to achieve UPF control, the reference current i_{qr-f} is obtained via Q_{ref} conforming to (6.12) and it is fixed to zero. On the other hand, one inner rapid dynamic loop is employed with the line current command and the SMA is chosen to follow the current reference. Also, the fundamental idea of a SMA approach is to propose a sliding surface in its command rule that will follow the required state variables toward its required references. So, defining the sliding surfaces (SS) as:

$$S_{d\text{-}grid} = i_{dr-f} - i_{d-f} \tag{6.16}$$

$$S_{q\text{-}grid} = i_{qr-f} - i_{q-f} \tag{6.17}$$

Using (6.9) and (6.10), we get:

$$\frac{dS_{d\text{-}grid}}{dt} = \frac{di_{dr-f}}{dt} - \frac{di_{d-f}}{dt} = \frac{di_{dr-f}}{dt} + \frac{R_f}{L_f} i_{d-f} - \omega i_{q-f} - \frac{1}{L_f} e_d + \frac{1}{L_f} U_0 \tag{6.18}$$

$$\frac{dS_{q\text{-}grid}}{dt} = \frac{di_{qr-f}}{dt} - \frac{di_{q-f}}{dt} = \frac{di_{qr-f}}{dt} + \frac{R_f}{L_f} i_{q-f} + \omega i_{d-f} - \frac{1}{L_f} e_q \tag{6.19}$$

If the sliding mode gets a place on the SS, so [20]:

$$S_{d\text{-}grid} = \frac{dS_{d\text{-}grid}}{dt} = 0 \tag{6.20}$$

$$S_{q\text{-}grid} = \frac{dS_{q\text{-}grid}}{dt} = 0 \tag{6.21}$$

The control laws can be obtained from (6.16) and (6.21) as

$$v_{dr\text{-}grid} = L_f \frac{di_{dr-f}}{dt} + R_f i_{d-f} - L_f \omega i_{q-f} + U_0 + L_f k_{d\text{-}grid} \operatorname{sgn}(S_{d\text{-}grid}) \tag{6.22}$$

$$v_{qr\text{-}grid} = R_f i_{q-f} + L_f \omega i_{d-f} + L_f k_{q\text{-}grid} \operatorname{sgn}(S_{q\text{-}grid}) \tag{6.23}$$

where $k_{d\text{-}grid} \succ 0$ and $k_{q\text{-}grid} \succ 0$. $\operatorname{sgn}(S_{dq\text{-}grid})$ is the following function [21,22]:

$$\operatorname{sgn}(S_{dq\text{-}grid}) = \begin{cases} 1\, S_{dq\text{-}grid} \succ \mu \\ \dfrac{S_{dq\text{-}grid}}{\mu} \mu \geq \left|S_{dq\text{-}grid}\right| \\ -1 - \mu \succ S_{dq\text{-}grid} \end{cases} \tag{6.24}$$

μ is the width of the boundary layer. The dynamic quality of the system will be reduced if μ is not chosen attentively.

Theorem 6.1: *The global asymptotical stability of the system, using SMA scheme with the control laws (6.22) and (6.23), is ensured and the tracking error can converge to zero asymptotically.*

Proof: The proof of Theorem 6.1 will be approved with the Lyapunov theory. So, selecting a Lyapunov function candidate as

$$\Gamma_{grid} = \frac{1}{2}S_{d\text{-}grid}^2 + \frac{1}{2}S_{q\text{-}grid}^2 \tag{6.25}$$

So,

$$\frac{d\Gamma_{grid}}{dt} = S_{d\text{-}grid}\frac{dS_{d\text{-}grid}}{dt} + S_{q\text{-}grid}\frac{dS_{q\text{-}grid}}{dt} \tag{6.26}$$

From (6.18) and (6.19), we obtain

$$\begin{aligned} S_{d\text{-}grid}\frac{dS_{d\text{-}grid}}{dt} &= S_{d\text{-}grid}\left[\frac{di_{dr\text{-}grid}}{dt} - \frac{1}{L_f}(L_f\omega i_{q-f} - R_f i_{d-f} + e_d - U_0)\right] \\ &= -k_{d\text{-}grid}S_{d\text{-}grid}\,\mathrm{sgn}(S_{d\text{-}grid}) \\ &\quad + \frac{S_{d\text{-}grid}}{L_f}\Big[L_f\frac{di_{dr-f}}{dt} - e_d + R_f i_{d-f} - L_f\omega i_{q-f} \\ &\quad + U_0 + k_{d\text{-}grid}L_f\,\mathrm{sgn}(S_{d\text{-}grid})\Big] \end{aligned} \tag{6.27}$$

$$\begin{aligned} S_{q\text{-}grid}\frac{dS_{q\text{-}grid}}{dt} &= S_{q\text{-}grid}\left[-\frac{di_{q-f}}{dt}\right] \\ &= -k_{q\text{-}grid}S_{q\text{-}grid}\mathrm{sgn}(S_{q\text{-}grid}) \\ &\quad + \frac{S_{q\text{-}grid}}{L_f}\left[-e_q + L_f\omega i_{d-f} + R_f i_{q-f} + k_{q\text{-}grid}L_f\,\mathrm{sgn}(S_{q\text{-}grid})\right] \end{aligned} \tag{6.28}$$

From (6.27) and (6.28), it can be concluded that

$$\begin{aligned} \frac{d\Gamma_{grid}}{dt} &= -k_{d\text{-}grid}S_{d\text{-}grid}\,\mathrm{sgn}(S_{d\text{-}grid}) \\ &+ \frac{S_{d\text{-}grid}}{L_f}\left[L_f\frac{di_{dr-f}}{dt} - L_f\omega i_{q-f} + R_f i_{d-f} - e_d + U_0 + k_{d\text{-}grid}L_f\,\mathrm{sgn}(S_{d\text{-}grid})\right] \\ &- k_{q\text{-}grid}S_{q\text{-}grid}\,\mathrm{sgn}(S_{q\text{-}grid}) \\ &+ \frac{S_{q\text{-}grid}}{L_f}\left[-e_q + L_f\omega i_{d-f} + R_f i_{q-f} + k_{q\text{-}grid}L_f\,\mathrm{sgn}(S_{q\text{-}grid})\right] \end{aligned} \tag{6.29}$$

Substituting (6.22) and (6.23) into (6.29) gives:

$$\frac{d\Gamma_{grid}}{dt} = -k_{d\text{-}grid}S_{d\text{-}grid}\,\mathrm{sgn}(S_{d\text{-}grid}) - k_{q\text{-}grid}S_{q\text{-}grid}\,\mathrm{sgn}(S_{q\text{-}grid}) \tag{6.30}$$

Consequently:

$$\frac{d\Gamma_{grid}}{dt} = -k_{d\text{-}grid}\left|S_{d\text{-}grid}\right| - k_{q\text{-}grid}\left|S_{q\text{-}grid}\right| \prec 0 \tag{6.31}$$

The global asymptotical stability of the current loop, using the SMA scheme, is ensured and the DC-bus voltage command tracking is achieved. The structure of the DC-link voltage and current regulators for the VSI is illustrated in Figure 6.2.

6.3 SMA for performance improvement of WPS fed by VSI

This section presents SMA for WPS based PMSG and connected to electrical network with VSI. The configuration of the system is illustrated is in Figure 6.3. The grid side converter (GSC) controls power flow to the grid. Moreover, it commands the DC bus voltage and the power factor. The rotor side converter (RSC) is in charge of implementing MPP.

6.3.1 Modeling description of the WECS

6.3.1.1 Wind turbine

The aerodynamic power extracted by the wind turbine (WT) is defined as [23,24]:

$$P_{turbine} = \frac{1}{2}\rho\pi R_t^2 C_p(\lambda_t, \beta_t) v_w^3 \pi \tag{6.32}$$

where ρ is the air density, R_t is the WT rotor radius, v_w is the wind speed, and $C_p(\lambda_t, \beta_t)$ is the efficiency of WT. λ_t and β_t are tip speed ratio (TSR) and pitch angle (PA), respectively. TSR is given as:

$$\lambda_t = \frac{\omega_t R_t}{v_w} \tag{6.33}$$

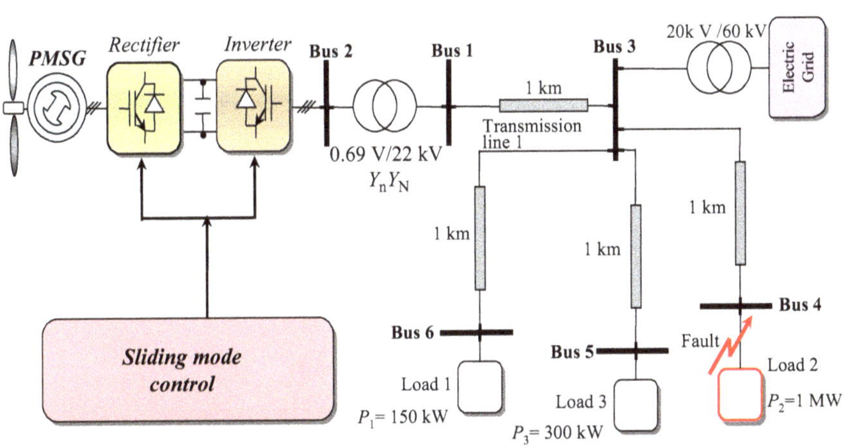

Figure 6.3 Schematic of proposed WPS

where ω_t is the WT velocity (in rad/sec). A generic equation is utilized to model $C_p(\lambda_t, \beta_t)$ as:

$$C_P = \frac{1}{2}\left(\frac{116}{\lambda_i} - 0.4\,\beta - 5\right)e^{-\left(\frac{21}{\lambda_i}\right)}$$
$$\frac{1}{\lambda_i} = \frac{1}{\lambda + 0.08\,\beta} - \frac{0.035}{\beta^3 + 1} \tag{6.34}$$

The aerodynamic torque is given by:

$$Ta = \frac{P_{Turbine}}{\omega_t} = \frac{1}{2}\rho\pi R_t^2 C_p(\lambda_t, \beta_t)v^3\frac{1}{\omega_t}$$

The WT model employed in this chapter has a maximum $C_p = C_{p-\max} = 0,41$ attained at an optimal $\lambda_t = \lambda_{t-optim} = 8,1$ and pitch angle $\beta = \beta_t = 0$. Therefore, according to (6.33), λ_t can be sustained at $\lambda_{t-optim}$ independently of v_w via changing the PMSG velocity. Hence MPP tracking is assured during diverse wind conditions. Accordingly, the maximum of $P_{turbine}$ is:

$$P_{turbine\_\max} = \frac{1}{2}\rho\pi R_t^2 C_{P\ \max}\left(\frac{R\omega_{t-optim}}{\lambda_{t-optim}}\right)^3 \tag{6.35}$$

Figure 6.4 illustrates the characteristics curve for the maximum power $P_{turbine\_\max}$ versus $\omega_{t-optim}$ with the maximum power curve in the MPPT control mode. Also, once the maximum rating of v_w is attained, the pitch angle is augmented to allow the WT to operate at low effectiveness. Thus, the PMSG speed will be at its nominal value. Figure 6.5 depicts the employed PA configuration. P_{PMSG} is the power of the PMSG.

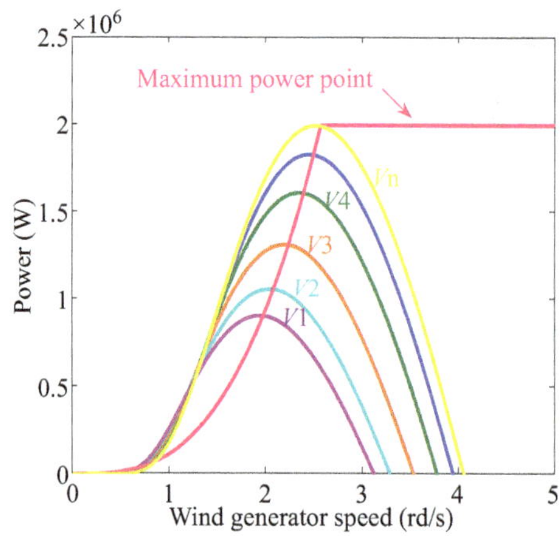

Figure 6.4 WT power versus WT speed

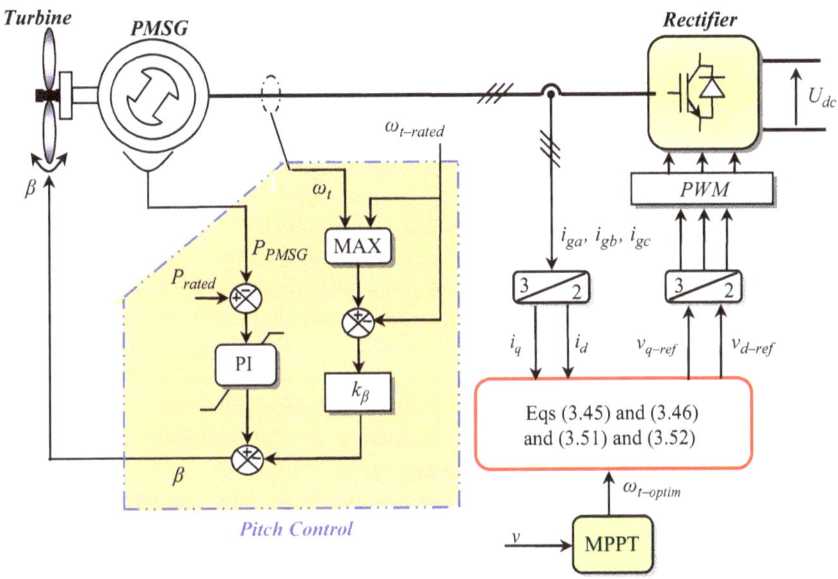

Figure 6.5 Configuration of the proposed control machine side converter

6.3.1.2 PMSG modeling

The implemented *dq* PMSG model is expressed below [25]:

$$\frac{di_q}{dt} = \frac{1}{L_s}(v_{gq} - R_g i_q - \omega_e L_s i_d - \omega_e \psi_f) \tag{6.36}$$

$$\frac{di_d}{dt} = \frac{1}{L_s}(v_{gd} - R_g i_d + \omega_e L_s i_q) \tag{6.37}$$

$$T_e = \frac{3}{2} p_n [\psi_f i_q] \tag{6.38}$$

$$J\frac{d\omega_m}{dt} = T_e - T_m - F\omega_t \tag{6.39}$$

The electrical rotating velocity of the generator:

$$\omega_e = p_n \omega_t \tag{6.40}$$

where

v_{gq}, v_{gd} is the stator voltage; i_q, i_d is the stator current; L_s the inductance of PMSG; R_g is the stator resistance; ψ_f is the flux induced by the magnet; p_n is the pole pair number; J is the total moment of inertia of the system (turbine-generator); F is the viscous friction coefficient; T_m is the torque developed by the WT.

6.3.2 SMA of the rectifier and MPP tracking approach

The MPP tracking of WPS is essential to enhance the effectiveness of the wind system. So, the rectifier controls the PMSG velocity to assure MPP from the WT. In addition, from (6.38) and (6.39), one can control the speed of the PMSG by regulating i_q. So, defining the SS as:

$$S_\omega = \omega_{t\_optim} - \omega_t \tag{6.41}$$

Thus:

$$\frac{dS_\omega}{dt} = \frac{d\omega_{t\_optim}}{dt} - \frac{d\omega_t}{dt} \tag{6.42}$$

Using (6.39), the time derivatives of S_ω can be calculated as:

$$\frac{dS_\omega}{dt} = \frac{d\omega_{t\_optim}}{dt} - \frac{1}{J}(T_e - T_m - F\omega_t) \tag{6.43}$$

If the SMA gets place on the SS, so [20]:

$$S_\omega = \frac{dS_\omega}{dt} = 0 \tag{6.44}$$

The control laws can be obtained from (6.38) and (6.43) as

$$i_{q-ref} = \frac{2}{3p_n\psi_f}\left(T_m + J\frac{d\omega_{t\_optim}}{dt} + F\omega_t + J\,k_\omega\,\mathrm{sgn}(S_\omega)\right) \tag{6.45}$$

$$i_{d-ref} = 0 \tag{6.46}$$

where $k_\omega \succ 0$.

On the other hand, to i_d and i_q to i_{d-ref} and i_{q-ref}, respectively, defining the SS as:

$$S_{d-PMSG} = i_{d-ref} - i_d \tag{6.47}$$

$$S_{q-PMSG} = i_{q-ref} - i_q \tag{6.48}$$

Consequently, using (6.36) and (6.37), the time derivatives of S_{d-PMSG} and S_{q-PMSG} can be calculated as:

$$\frac{dS_{d-PMSG}}{dt} = \frac{di_{d-ref}}{dt} - \frac{di_d}{dt} = -\frac{1}{L_s}(v_{gd} - R_g i_d + L_s\omega_e i_q) \tag{6.49}$$

$$\begin{aligned}\frac{dS_{q-PMSG}}{dt} &= \frac{di_{q-ref}}{dt} - \frac{di_q}{dt}\\ &= \frac{di_{qr}}{dt} - \frac{1}{L_s}(-R_g i_q - L_s\omega_e i_d - \omega_e\psi_f + v_{gq})\end{aligned} \tag{6.50}$$

If the SMA mode gets place on the SS, the control laws can be obtained from (6.47) and (6.50) as

$$v_{d-ref} = R_g i_d - L_s \omega_e i_q + L_s k_d \operatorname{sgn}(S_{d-PMSG}) \tag{6.51}$$

$$v_{q-ref} = L_s \frac{di_{q-ref}}{dt} + \omega_e \psi_f + L_s \omega_e i_d + R_g i_q + L_s k_q \operatorname{sgn}(S_{q-PMSG}) \tag{6.52}$$

k_q and k_d are positive constants.

Theorem 6.2 *The global asymptotical system stability, using SMA scheme via Equation laws (6.45)–(6.46) and (6.51)–(6.52), is ensured and the tracking errors can converge asymptotically towards zero.*

Proof: The proof of Theorem 6.2 will be approved with the theory of Lyapunov. So, selecting a function candidate as

$$\Gamma_{PMSG} = \frac{1}{2} S_\omega^2 + \frac{1}{2} S_{d-PMSG}^2 + \frac{1}{2} S_{q-PMSG}^2 \tag{6.53}$$

Taking the derivative of (6.53) gives

$$\frac{d\Gamma_{PMSG}}{dt} = S_\omega \frac{dS_\omega}{dt} + S_{d-PMSG} \frac{dS_{d-PMSG}}{dt} + S_{q-PMSG} \frac{dS_{q-PMSG}}{dt} \tag{6.54}$$

From (6.43), (6.49), and (6.50), we obtain

$$\begin{aligned} S_\omega \frac{dS_\omega}{dt} &= S_\omega \frac{d\omega_{t\_optim}}{dt} - \frac{S_\omega}{J}(T_e - T_m - F\omega_t) \\ &= -k_\omega S_\omega \operatorname{sgn}(S_\omega) + \frac{S_\omega}{J}\left(-\frac{3}{2} p_n \psi_f i_q + T_m + Jk_\omega \operatorname{sgn}(S_\omega)\right. \\ &\quad \left. + F\omega_t + J\frac{d\omega_{t\_optim}}{dt}\right) \end{aligned} \tag{6.55}$$

$$\begin{aligned} S_{d-PMSG} \frac{dS_{d-PMSG}}{dt} &= S_{d-PMSG}\left[-\frac{1}{L_s}(v_{gd} - R_g i_d + L_s \omega_e i_q)\right] \\ &= -k_d S_{d-PMSG} \operatorname{sgn}(S_{d-PMSG}) \\ &\quad + \frac{S_{d-PMSG}}{L_s}\left[-L_s \omega_e i_q + R_g i_d - v_{gd} + L_s k_d \operatorname{sgn}(S_{d-PMSG})\right] \end{aligned} \tag{6.56}$$

$$\begin{aligned} S_{q-PMSG} \frac{dS_{q-PMSG}}{dt} &= S_{q-PMSG}\left[\frac{di_{q-ref}}{dt} - \frac{di_q}{dt}\right] \\ &= -k_q S_{q-PMSG} \operatorname{sgn}(S_{q-PMSG}) \\ &\quad + \frac{S_{q-PMSG}}{L_s}\left[L_s \frac{di_{q-ref}}{dt} + \omega_e \psi_f + R_g i_q + L_s \omega_e i_d - v_{gq}\right. \\ &\quad \left. + L_s k_q \operatorname{sgn}(S_{q-PMSG})\right] \end{aligned} \tag{6.57}$$

Using (6.55) and (6.57), we can rewrite (6.54) as:

$$\begin{aligned}\frac{d\Gamma_{PMSG}}{dt} &= \frac{S_\omega}{J}\left(T_m + F\omega_m - \frac{3}{2}p_n\psi_f i_q + Jk_\omega \operatorname{sgn}(S_\omega) + J\frac{d\omega_{m\_opt}}{dt}\right) - k_\omega S_\omega \operatorname{sgn}(S_\omega)\\ &- k_d S_{d-PMSG} \operatorname{sgn}(S_{d-PMSG}) + \frac{S_{d-PMSG}}{L_s}\left[-v_{gd} + R_g i_d - L_s\omega_e i_q + L_s k_d \operatorname{sgn}(S_{d-PMSG})\right]\\ &- k_q S_{q-PMSG} \operatorname{sgn}(S_{q-PMSG}) + \frac{S_{q-PMSG}}{L_s}\left[L_s\frac{di_{q-ref}}{dt} + \omega_e\psi_f + R_g i_q + L_s\omega_e i_d - v_{gq} + L_s k_q \operatorname{sgn}(S_{q-PMSG})\right]\end{aligned} \tag{6.58}$$

Substituting (6.45)–(6.46) and (6.51)–(6.52) into the above equation gives:

$$\begin{aligned}\frac{d\Gamma_{PMSG}}{dt} &= -k_d S_{d-PMSG} \operatorname{sgn}(S_{d-PMSG}) - k_\omega S_\omega \operatorname{sgn}(S_\omega)\\ &- k_q S_{q-PMSG} \operatorname{sgn}(S_{q-PMSG})\end{aligned} \tag{6.59}$$

Consequently:

$$\frac{d\Gamma_{PMSG}}{dt} = -k_\omega|S_\omega| - k_d|S_{d-PMSG}| - k_q|S_{q-PMSG}| \prec 0 \tag{6.60}$$

The global asymptotical stability of the PMSG speed, using SMA scheme, is ensured and the MPPT approach is achieved. The structure of the speed and AC/DC converter is depicted in Figure 6.5.

6.4 Simulation and evaluation of performance

In this section, the SMA has been simulated via MATLAB®/Simulink® to evaluate the performance of the proposed approach for the configuration depicted in Figure 6.3 which is based on Figures 6.2 and 6.5. The parameters of the simulation are presented in the Appendix. $U_{dc-ref} = 1500\ V$ and the grid frequency value is 50 Hz. Wind velocity profile is depicted in Figure 6.6(a). $v_n = 12.4$ m/s is the rated wind velocity. Figure 6.6(c) shows the evolution of the power coefficient C_p. At first, the generation of maximum power is completely possible because the turbine operates at optimal $\lambda_{t-optim}$ that will guarantee $C_{p-\max}$. It can be seen that the value C_p is fixed at $C_{p-\max}$ despite all the deviations in the wind velocity. As displayed in Figure 6.6(c), the SMA can command C_p to be sustained at its peak value. From Figure 6.6(b), it is noted that the pitch angle control has a considerable impact on high wind speed. Indeed, it can be successfully employed and can modify the C_p. Figure 6.6(d) depicts the tracking performance of the speed PMSG. The speed regulator controls the velocity of the generator in reaction to varies in the wind speed. The generated reference velocity $\omega_{t-optim}$ versus the actual speed of the PMSG ω_t is presented. It is observed from Figure 6.8(e) that the regulator guarantees an efficient tracking of the required trajectory. So, the speed of the PMSG tracks the reference velocity, demonstrating the rectifier capability to employ the MPPT approach. The extracted power is proportional to the wind speed

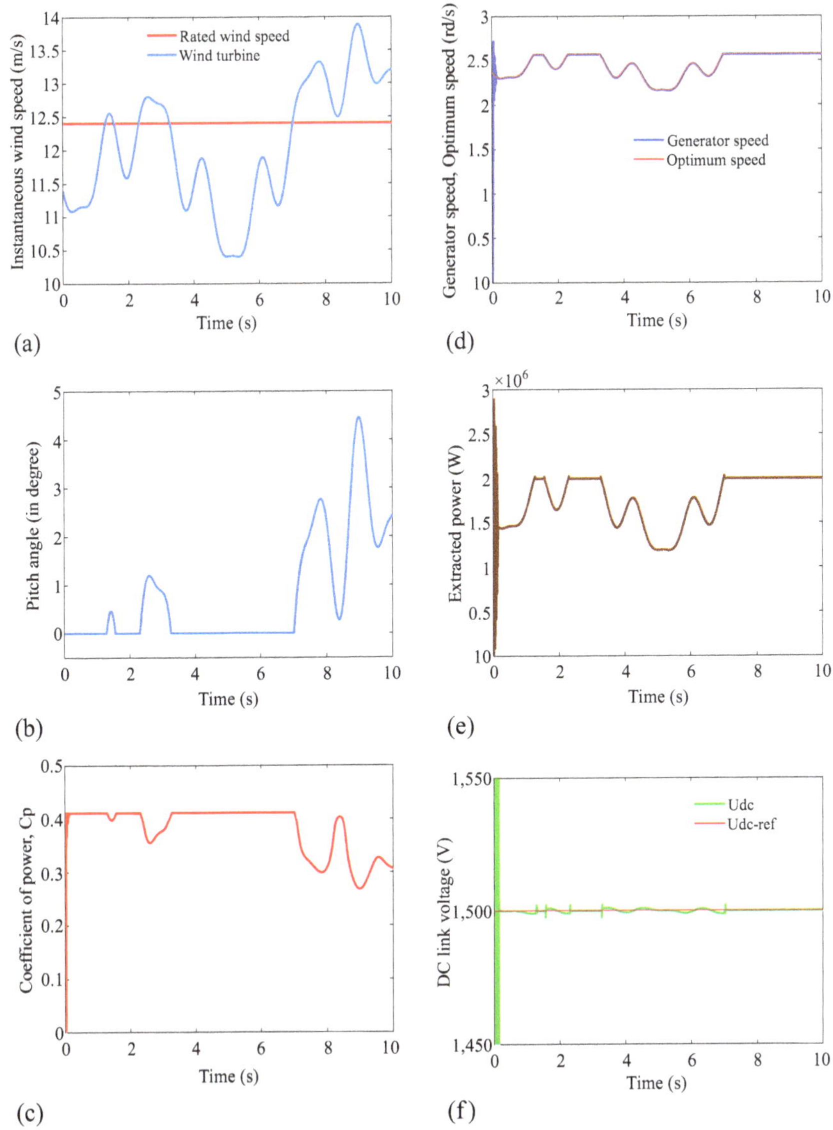

Figure 6.6 Waveforms of WECS characteristics with SM approach. (a) Wind speed profile (m/s), (b) pitch angle β, (c) variation of coefficient of power. (d) Speed of PMSGs, (e) extracted power of WECS, (f) DC link voltage (V).

deviation as depicted in Figure 6.6(e). Figure 6.6(f) depicts DC link voltage which is exactly close to $U_{bus-ref}$ during diverse wind speed. From Figure 6.7, it is noted that the grid voltage is in phase with the current since $Q_{ref} = 0$ VAR. At $t = 4$ s, the voltage symmetric fault happens in the utility-grid at Bus 4

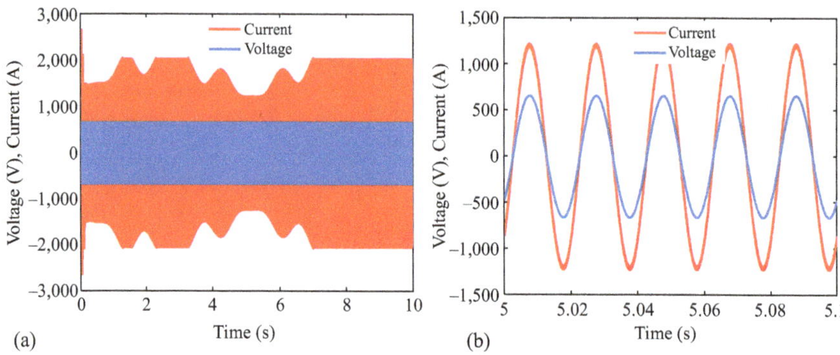

Figure 6.7 Three phase current and voltage

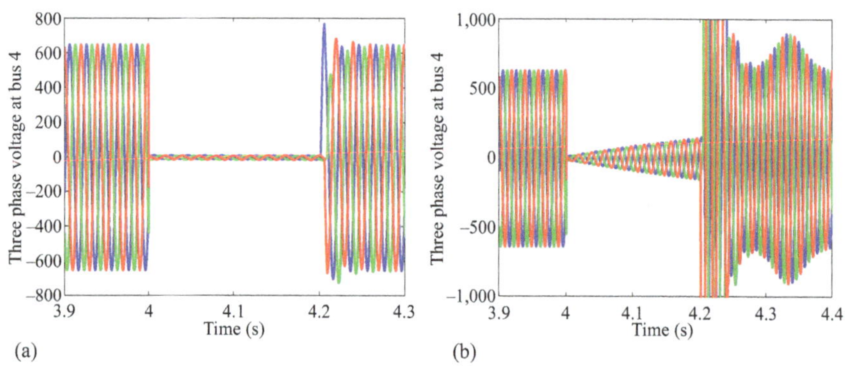

Figure 6.8 Zoom-in view of three phase voltage at Bus 4: (a) Sliding mode control and (b) vector control approach

(Figure 6.3). After 200 ms, it returned to normal, as depicted in Figure 6.8. The performance of the proposed SMC is compared to the vector control (VC) and displayed in Figures 6.8 and 6.9. One clearly concludes that the proposed SMA, in contrast with the VC, offers a smoother steady-state response and ensures small oscillation voltage. Also, it gives faster transient response with less fluctuation under fault conditions. Figure 6.9 shows the frequency performance with SMC and VC approach. From the presented results, maximum frequency deviations are reduced from 2.1 Hz to less than 0.22 Hz with SM controllers. As a result, the frequency deviations are significantly diminishing during the voltage fault and it is very apparent that the system is few responsive to power network faults with SM approach. The harmonic content of the current injected in the grid was evaluated by the calculation of the total harmonic distortion (THD) using fast Fourier transform (FFT) analysis. The FFT spectrum of the electrical network side current is depicted in Figure 6.10. Indeed, the improvement of the grid current THD is regarding 0.11% for the proposed method and 0.17% for the VC approach.

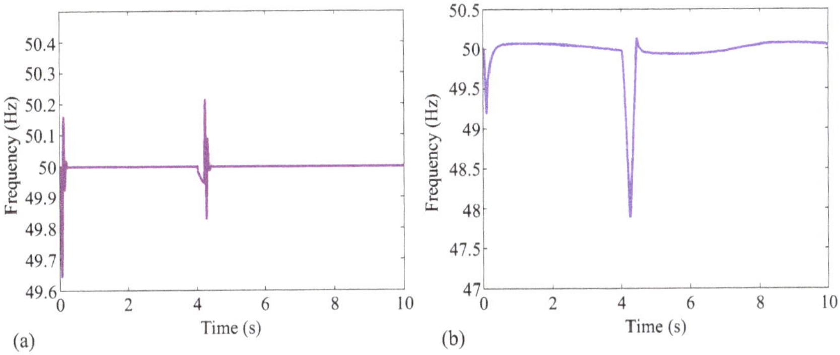

Figure 6.9 Frequency response of the WEC under grid faults. (a) Sliding mode control and (b) vector control approach.

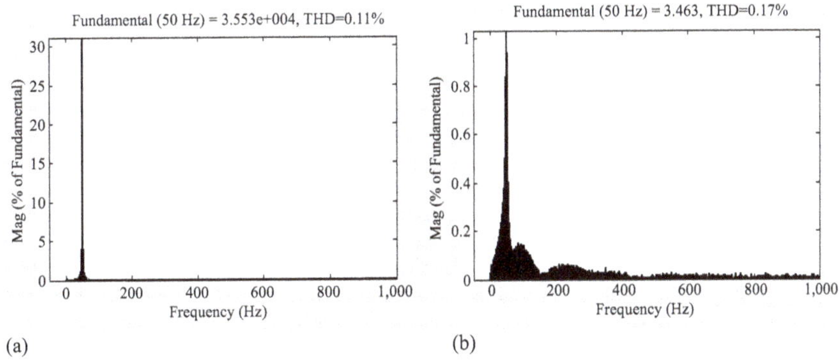

Figure 6.10 Frequency spectra of current injected in the grid. (a) Sliding mode control and (b) vector control approach.

6.5 Conclusions

In this chapter, a control scheme-based SMA for a VSI was developed. A VSI is employed to connect the WPS to the electrical grid and to process the energy generated by this system. VSI control can guarantee the stability of DC bus voltage and the synchronization of the power, produced by the system, with the grid. Also, it adjusts the powers exchanged with the power network and guarantees unity power factor (UPF). The comparative results and analysis for the presented SMA strategy and VC are presented for different grid voltage conditions to demonstrate the excellent performances of the proposed method such as smoother transient responses and less fluctuation under fault conditions.

Appendix

Table 6.1 Parameters of the power synchronous generator

Parameter	Value
P_r-rated power	2 MW
ω_m-rated mechanical speed	2.57 rad/s
R stator resistance	0.008 Ω
L_s stator d-axis inductance	0.0003 H
ψ_f permanent magnet flux	3.86 wb
p_n pole pairs	60

Table 6.2 Parameters of the turbine

Parameter	Value
ρ the air density	1.08 kg/m^3
A area swept by blades	4,775.94 m^2
v_n base wind speed	12.4 m/s

References

[1] L. Zheng, F. Jiang, J. Song, Y. Gao, and M. Tian, 'A discrete time repetitive sliding mode control for voltage source inverters', *IEEE Journal of Emerging and Selected Topics in Power Electronics*. vol. 6, no. 3, pp. 1553–1566, 2018.

[2] M.S. Rafaq, S.A. Qasem Mohammed, H.H. Choi, and J. Jung, 'An improved sliding mode control technique to mitigate mismatched parameter uncertainties of three-phase voltage source inverters'. *IEEE Access*. vol. 8, pp. 81932–81942, 2020.

[3] H. Shariatpanah, R. Fadaeinedjad, and M. Rashidinejad, 'A new model for PMSG-based wind turbine with yaw control', *IEEE Transactions On Energy Conversion*. vol. 28, no. 4, pp. 929–937, 2013.

[4] Shuhui Li, T. A. Haskew, R. P. Swatloski, and W. Gathings, 'Optimal and direct-current vector control of direct-driven PMSG wind turbines', *IEEE Transactions on Power Electronics*. vol. 27, no. 5, pp. 2325–2337, 2012.

[5] S. Alepuz, A. Calle, S. Busquets-Monge, S. Kouro, and B. Wu, 'Use of stored energy in PMSG rotor inertia for low-voltage ride-through in back-to-back NPC converter-based wind power systems', *IEEE Transactions on Industrial Electronics*. vol. 60, no. 5, pp. 1787–1796, 2013.

[6] Y. Zhang, J. Gao, and C. Qu, 'Relationship between two direct power control methods for PWM rectifiers under unbalanced network,' *IEEE Transactions on Power Electronics*. vol. 32, no. 5, pp. 4084–4094, 2017.

[7] T. Noguchi, H. Tomiki, S. Kondo, and I. Takahashi, 'Direct power control of PWM converter without power-source voltage sensors,' *IEEE Transactions on Industry Applications*. vol. 34, no. 3, pp. 473–479, 1998.
[8] M. Malinowski, M. Jasiński, and M. P. Kazmierkowski, 'Simple direct power control of three-phase PWM rectifier using space-vector modulation (DPC-SVM),' *IEEE Transactions on Industrial Electronics*. vol. 51, no. 2, pp. 447–454, 2004.
[9] F. Evran, 'Plug-in repetitive control of single-phase grid-connected inverter for AC module applications,' *IET Power Electronics*. vol. 10, no. 1, pp. 47–58, 2017.
[10] F. Xiao, L. Dong, S. F. Khahro, X. Huang, and X. Liao, 'A smooth LVRT control strategy for single-phase two-stage grid-connected PV inverters,' *Journal of Power Electronics*. vol. 15, no. 3, *pp. 806–818*, 2015.
[11] R. Errouissi and A. Al-Durra, 'Disturbance observer-based control for dual-stage grid-tied photovoltaic system under unbalanced grid voltages,' *IEEE Transactions on Industrial Electronics*. vol. 6, no. 11, pp. 8925–8936.
[12] A. Abrishamifar, A. Ahmad, and M. Mohamadian, 'Fixed switching frequency sliding mode control for single-phase unipolar inverters,' *IEEE Transactions on Power Electronics*. vol. 27, no. 5, pp. 2507–2514, 2012.
[13] R. Guzman, L. G. de Vicuña, M. Castilla, J. Miret, and H. Martín, 'Variable structure control in natural frame for three-phase grid-connected inverters with LCL filter,' *IEEE Transactions on Power Electronics*. vol. 33, no. 5, pp. 4512–4522, 2017.
[14] A. Merabet, L. Labib, A. M. Y. M. Ghias, C. Ghenai, and T. Salameh, 'Robust feedback linearizing control with sliding mode compensation for a grid-connected photovoltaic inverter system under unbalanced grid voltages,' *IEEE Journal of Photovoltaics*. vol. 7, no. 3, pp. 828–838, 2017.
[15] Y. Zhu and J. Fei, 'Disturbance observer based fuzzy sliding mode control of PV grid connected inverter,' *IEEE Access*. vol. 6, pp. 21202–21211, 2018.
[16] S. Oucheriah and L. Guo, 'PWM-based adaptive sliding-mode control for boost DC–DC converters,' *IEEE Transactions on Industrial Electronics*. vol. 60, no. 8, pp. 3291–3294, 2013.
[17] R. P. Vieira, L. Tom′ e Martins, J. Rodrigo Massing, and M. Stefanello, 'Sliding mode controller in a multi–loop framework for a grid–connected VSI with LCL filter,' *IEEE Transactions on Industrial Electronics*. vol. 65, no. 6, pp. 4714–4723, 2018.
[18] X. Zheng, Y. Feng, F. Han, and X. Yu, 'Integral-type terminal sliding-mode control for grid-side converter in wind energy conversion systems,' *IEEE Transactions on Industrial Electronics*. vol. 66, no. 5, pp. 3702–3711, 2019, doi:10.1109/tie.2018.2851959.
[19] H. Tingting, D. D. Lu, L. Li, J. Zhang, L. Zheng, and J. Zhu, 'Model predictive sliding-mode control for three-phase AC/DC converters,' *IEEE Transactions on Power Electronics*. vol. 33, no. 10, pp. 8982–8993, 2018.
[20] C. Evangelista, P. Puleston, F. Valenciaga, and L. M. Fridman, 'Lyapunov-designed super-twisting sliding mode control for wind energy conversion

optimization,' *IEEE Transactions on Industrial Electronics*. vol. 60, no. 2, pp. 538–545, 2013.

[21] L. Xiao, S. Huang, L. Zheng, Q. Xu, and K. Huang, 'Sliding mode SVM-DPC for grid-side converter of D-PMSG under asymmetrical faults,' In *IEEE International Conference on Electrical Machines and Systems (ICEMS)*, pp. 1–6, 2011.

[22] Y. Errami, M. Hilal, M. Benchagra, M. Ouassaid, and M. Maaroufi, 'Nonlinear control of MPPT and grid connected for variable speed wind energy conversion system based on the PMSG,' *Journal of Theoretical and Applied Information Technology*. vol. 39, no. 2, pp. 204–217, 2012.

[23] Y. Errami, M. Ouassaid, and M. Maaroufi, 'Variable structure control for permanent magnet synchronous generator based wind energy conversion system operating under different grid conditions,' In *2014 2nd World Conference on Complex Systems, WCCS 2014*, 2015, pp. 340–345, 7060996.

[24] M. Hilal, M. Benchagra, Y. Errami, M. Maaroufi, and M. Ouassaid, 'Maximum power tracking of wind turbine based on doubly fed induction generator,' *International Review on Modelling and Simulations*. vol. 4, no. 5, pp. 2255–2263, 2011.

[25] Y. Errami, M. Ouassaid, and M. Maaroufi, 'Control scheme and power maximisation of permanent magnet synchronous generator wind farm connected to the electric network,' *International Journal of Systems, Control and Communications*. vol. 5, no. 3–4, pp. 214–230, 2013.

Chapter 7

Sliding-mode control of a three-level NPC grid-connected inverter

Amine El Fathi[1], Hamid El-Moumen[1], Saloua Yahyaoui[2], Nabil El Akchioui[1] and Mohamed Bendaoud[3]

In this chapter, a super-twisting SMC (STSMC) will be used to control a three-phase natural clamped point (NPC) inverter with LCL filter. The main objective is to ensure a regulated output AC current with a low Total Harmonic Distortion (THD) value. MATLAB®/Simulink® is used to show the effectiveness of the STSMC strategy. The benefits of the proposed control strategy can be summarized as compliance with the standards imposed by the electrical grid managers without the use of onerous filters at the output of the NPC inverter. Furthermore, the proposed control strategy exhibits an accurate tracking response with low THD of the injected currents while ensuring grid voltage synchronization.

7.1 Introduction

Grid-connected inverters have received a lot of attention in recent years [1,2]. This is due to the increased use of clean energy sources whether in standalone systems or grid-connected systems. The latter requires compliance with certain standards concerning the quality of the energy introduced into the electrical grid [3]. Indeed, some standards have been specially developed to guarantee a better quality of the grid voltage whether in amplitude, frequency, and phase. In the same context, the quality of the current controlled by a grid-connected inverter is also of concern. Indeed, a better control strategy for the grid-connected inverter is essential in order to respect the international standards related to the quality of the energy and the requirements imposed by the managers of the electrical grid. IEC 61000-2-2 Standard and IEC 61000-2-12 standard, imposed by the International

[1]LRDSI Laboratory, Faculty of Sciences and Technology of Al Hoceima, Abdelmalek Essaadi University, Morocco
[2]Electrical Engineering and Maintenance Laboratory, Higher School of Technology, Mohamed I University, Morocco
[3]Science and Technology for the Engineer Laboratory (LaSTI), Sultan Moulay Slimane University, Morocco

Electrotechnical Commission (IEC), define the compatibility margins during the transmission of signals at the low and medium voltage on the electrical grid. Furthermore, these two IEC standards, as well as IEEE 519 and EN 50160, set the total harmonic voltage distortion limit at 8%. They also specify for the voltage limits that they must not exceed for the first 50 individual harmonics. Interconnection specifications between photovoltaic (PV) systems and the electrical grid are defined by IEC 61727 standard [4]. In the latter, the THD factor for the current is limited to 5% while it is limited to 2% for the THD of the voltage. In addition, a maximum value of 1% for the individual harmonics of the voltage has been imposed. Generally, to reduce the influence of harmonic current components, the inverter is linked to the grid network through an L, RL, or LCL filter. The purpose of these filters is the attenuation of harmonics due to switching when controlling the inverter switches. In this context, a grid-connected inverter followed by an LCL filter exhibits better attenuation capability with lower inductance compared to L and LC filters [5,6]. It can be seen that the LCL and L filters exhibit comparable filtering properties at low frequencies and high frequencies. However, at the resonant frequency, the LCL filter generates a significant overshoot, resulting in amplitude amplification of the high-order harmonics of the current injected into the grid. As a result, if the disadvantages of LCL filters can be avoided while maintaining their benefits, grid-connected inverter performance will improve [7]. Several grid-connected inverter control strategies were used to meet the requirements of the electrical grid to which it would be connected. Among these control strategies, we find strategies based on linear, nonlinear, predictive, adaptive, and intelligent controllers [8]. A predictive controller predicts the behavior of parameters that must be controlled in the future. It is well known for its capability to manage the system's non-linearities. Predictive controllers can control current with low harmonic distortion due to their fast dynamic response. However, it is difficult to implement because it demands a large number of calculations and must be connected to an adequate electrical load [9]. The adaptive controllers can adapt the control actions in response to the system's operating parameters [10,11]. The major drawback of this controller is its large computational complexity [12].

Non-linear controllers have better performance and a higher dynamic response than linear controllers. Non-linear techniques outperform the other approaches in terms of their performance against parameter variation [13]. Sliding mode control (SMC) has been recognized to be among the most suitable nonlinear control approaches. In addition to their very fast transient response, they are easy to implement which means a lower cost. Moreover, they are insensitive to variations/disturbances of the variables of the studied system [14,15]. Despite all these strengths, SMC controllers have some drawbacks. For example, choosing the right sliding surface is a very complicated task. However, the main drawback of the SMC is manifested in the phenomenon of chattering which is observed during the reference tracking. In fact, chatter occurs due to the occurrence of a discontinuity in the control input [16,17]. In order to reduce the chattering effect, some interesting research works have been done using the reaching law proposed by Gao *et al.* in [18]. The SMC developed in these studies, however, requires a derivative operation [19]. It also requires a high accurate value of the line/filter resistor and inductor. The main goal of this work is to propose a method for controlling grid-connected three-level NPC inverters with a super twisting sliding mode-based controller (STSMC) and an LCL filter. In fact, the STSMC is a robust and efficient control technique that has been widely used in the control of three-phase grid-connected inverters [20]. The technique is based on the combination of the SMC and the super-twisting algorithm, which results in a robust and fast control system [21]. In this work, the control strategy does not require the derivative function and does not depend on LCL parameters. On the other hand, the PWM signals can be obtained by using the control inputs compared to high frequency triangular carrier signals leading to a fixed switching frequency which is appreciated in practice instead of time-varying switching frequency. As previously stated, the error between the reference current and the injected current into the grid should be zero. Furthermore, it should have low chattering and a constant switching frequency. Additionally, the control strategy should be simple to implement. Robustness is another critical criterion to consider. In fact, sudden changes in the grid parameters are considered in order to assess the steady-state evolution of the injected current. MATLAB/Simulink is used to show the effectiveness of the STSMC strategy.

7.2 Three-phase grid-connected NPC inverter

The growing use of multilevel inverters in renewable energy conversion applications is principally due to their natural ability to reduce harmonics, their low switching frequency operation, and ability to provide high power ratings, resulting in strong penetration of renewable energy resources such as solar and wind [22,23]. The most common topologies in multilevel inverters are neutral-point-clamped (NPC) [24], flying-capacitor [25], and cascaded H-bridge [26]. Due to the smaller size DC buses for three-leg/phase topologies, the three-level NPC inverter is the most appealing topology for renewable energy conversion systems [27].

Neutral point clamped (NPC) inverters are multi-level inverters. They are characterized by the use of clamping diodes in order to guarantee voltage sharing between the switches to be controlled. NPC inverters were introduced simultaneously by Baker [28] and Nabae *et al.* in 1981 [29]. Today, they remain the most widely used multilevel converter topology, as they offer reduced complexity with very attractive performances. Figure 7.1 depicts the circuit topology of three-level NPC grid-connected inverter.

Table 7.1 shows the output voltage between phase a, and the midpoint between the two capacitors for each combination of switching states. The same values are obtained for phases b and c. The zero-voltage level achieved by this structure increases the number of output voltage levels (3 levels), resulting in an output voltage with low harmonic content.

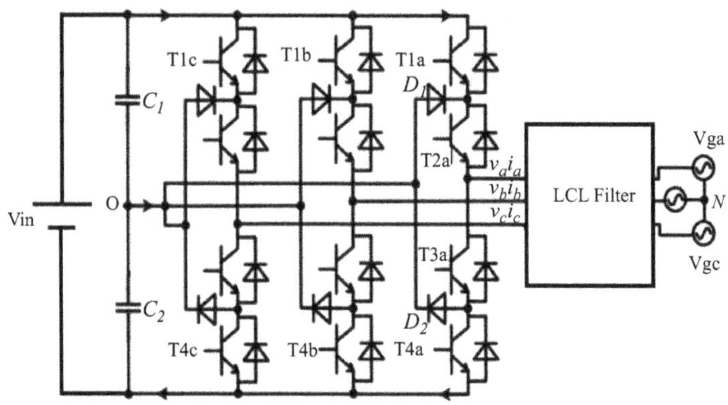

Figure 7.1 The circuit configuration of a three-level NPC grid-connected inverter with LCL filter

Table 7.1 Control input, switching states, and output voltage of a three-level NPC grid-connected inverter

Control input	Switching states				Output voltage
U_a	T_{1a}	T_{2a}	T_{3a}	T_{4a}	V_{ao}
+1	1	1	0	0	$+V_{in}/2$
0	0	1	1	0	0
−1	0	0	1	1	$-V_{in}/2$

In the stationary reference frame, the dynamic model of the system in Figure 7.1 is shown as follows:

$$L\frac{di_G}{dt} = U_G V_{in} - V_{g,G} - Ri_G \tag{7.1}$$

$$G = (a, b, c), U_G = \begin{bmatrix} u_a \\ u_b \\ u_c \end{bmatrix}, \quad V_{g,G} = \begin{bmatrix} v_{g,a} \\ v_{g,b} \\ v_{g,c} \end{bmatrix} \text{ and } i_G = \begin{bmatrix} i_a \\ i_b \\ i_c \end{bmatrix} \tag{7.2}$$

where V_G are the inverter output voltages, i_G is the grid currents, $V_{g,G}$ are the grid voltages, U_G is the switching function, V_{in} is the DC bus voltage, and L and R are the line inductor and the resistor, respectively. It should be noted that in (7.1), the diode conduction and switching losses are neglected. As mentioned above, LCL and L filters show comparable filtering properties at low frequencies and high frequencies, that is why only L is taken into consideration during the design of the control.

In order to make the modelling and the control design simpler, (7.1) is transformed into a dq frame as shown in (7.3). This latter is obtained by using the transformation from a stationary to a synchronous rotating frame. The matrix in

Equation (7.5) is used in this case:

$$\begin{cases} L\dfrac{di_d}{dt} = \dfrac{V_{in}}{2}U_d - v_{g,d} - Ri_d + wLi_q \\ L\dfrac{di_q}{dt} = \dfrac{V_{in}}{2}U_q - v_{g,q} - Ri_q - wLi_d \end{cases} \tag{7.3}$$

where

$$w = \frac{d\theta}{dt}, \quad \begin{bmatrix} i_d \\ i_q \end{bmatrix} = \frac{2}{3}Ti_G, \quad \begin{bmatrix} v_d \\ v_q \end{bmatrix} = \frac{2}{3}Tv_G \text{ and } \begin{bmatrix} u_d \\ u_q \end{bmatrix} = \frac{2}{3}TU_G \tag{7.4}$$

$$T = \sqrt{\frac{2}{3}}\begin{bmatrix} \cos\theta & \cos(\theta - 2\pi/3) & \cos(\theta + 2\pi/3) \\ -\sin\theta & -\sin(\theta - 2\pi/3) & -\sin(\theta + 2\pi/3) \end{bmatrix} \tag{7.5}$$

The control goal is to make sure that the dq currents (i_d and i_q) track their references, i_dref and i_qref, such as:

$$i_d \to i_d^{ref} \text{and} i_q \to i_q^{ref} \tag{7.6}$$

7.3 Reaching law in SMC

The SMC's goal is to provide high dynamic tracking performance for inverter output currents. First, the sliding surface should be selected based on the desired dynamics. After defining the sliding surface, the reaching law that guarantees stability of the system on the surface is developed. Finally, the generated SMC law is used as a control signal where it will be compared with triangle carriers in order to control the different switches of the NPC inverter.

7.3.1 Sliding surface design

A sliding mode current control's fundamental principle is to create a specific sliding surface in its control law that will follow the required state variables in the direction of the desired references. In order to achieve this, the synchronous reference frame introduces two sliding surfaces: δ_d for managing the direct current and δ_q for regulating the indirect one. The reference currents for the *d*-axis and the *q*-axis are, respectively, i_d^{ref} and i_q^{ref}, which the control system should track the evolution of (7.7):

$$\begin{cases} \delta_d = i_d - i_d^{ref} \\ \delta_q = i_q - i_q^{ref} \end{cases} \tag{7.7}$$

where i_d^{ref} and i_q^{ref} are i_d and i_q references, respectively.

The goal of SMC is to obtain $\delta = 0$ as represented in (7.8). The performance of the SMC is evaluated by monitoring this parameter, especially in the presence of any disturbances:

$$\delta = \begin{bmatrix} \delta_d \\ \delta_q \end{bmatrix} = 0 \tag{7.8}$$

Then, an equivalent control law U_{eq} that meets the condition $\dot{\delta} = 0$ should be developed. Consequently, (7.9) is deduced:

$$\dot{\delta} = \begin{bmatrix} \dot{\delta}_d \\ \dot{\delta}_q \end{bmatrix} = \begin{bmatrix} \frac{di_d}{dt} - \frac{di_d^{ref}}{dt} \\ \frac{di_q}{dt} - \frac{di_q^{ref}}{dt} \end{bmatrix} = 0 \tag{7.9}$$

The major disadvantage of the SMC is the chattering problem caused by the control law discontinuity. In the SMC design, reducing system chattering remains a challenge. Gao *et al.* proposed a comprehensive definition for the reaching law that reduces system chattering [18]. It can be expressed as an equation by (7.10):

$$\dot{\delta} = \begin{bmatrix} -\varepsilon_d \mathrm{sgn}(\delta_d) - K_d\delta_d \\ -\varepsilon_q \mathrm{sgn}(\delta_q) - K_q\delta_q \end{bmatrix} \tag{7.10}$$

where $\varepsilon_d > 0$, $\varepsilon_q > 0$, $k_d > 0$, and $K_q > 0$.

The second step of the SMC is to calculate the input control that ensures the stability of the system. The condition of system stability can be obtained if $\delta_d\dot{\delta}_d < 0$ and $\delta_q\dot{\delta}_q < 0$. Equations (7.9) = (7.10) are used to calculate the input control. Since $i_{d,q}^{ref}$ is constant, its derivative is equal to 0. Therefore, (7.11) is determined by computing the derivative of (7.7) and integrating then (7.3) and (7.10):

$$\begin{cases} U_d = \frac{2}{V_{in}}\left(-L(\varepsilon_d \mathrm{sgn}(\delta_d) + K_d\delta_d) + \quad V_{g,d} - LWi_q\right) \\ U_q = \frac{2}{V_{in}}\left(-L\left(\varepsilon_q \mathrm{sgn}(\delta_q) + K_q\delta_q\right) + V_{g,q} + LWi_d\right) \end{cases} \tag{7.11}$$

7.4 Super twisting SMC

7.4.1 Control design

The derivative of (7.7) leads to (7.12):

$$\begin{cases} \dot{\delta}_d = \frac{-1}{L}\left(V_{g,d} + Ri_d - LWi_q\right) + \frac{V_{in}}{2L}U_d \\ \dot{\delta}_q = \frac{-1}{L}\left(V_{g,q} + Ri_q + LWi_d\right) + \frac{V_{in}}{2L}U_q \end{cases} \tag{7.12}$$

As represented in (7.13), (7.12) can be written in the following form:

$$\dot{\delta} = C + DU_{dq}y = \delta \tag{7.13}$$

where δ is the sliding surface vector, U is the control input, y is the controlled output, C and D are given by (7.14):

$$\delta = \begin{bmatrix} \delta_d \\ \delta_q \end{bmatrix},\ C = \frac{-1}{L}\begin{bmatrix} V_{g,d} + Ri_d + LWi_q \\ V_{g,q} + Ri_q + LWi_d \end{bmatrix} \text{ and } D = \frac{V_{in}}{2L} \quad U_{dq} = \begin{bmatrix} U_d \\ U_q \end{bmatrix} \tag{7.14}$$

The aim of the control is to derive U_{dq} such that the error variables converge to $y = dy/dt = 0$ in finite time. The major drawback of conventional SMC is chattering. In order to avoid this, an STSMC is proposed for first-order systems which retains the attractive properties of SMC. Therefore, in STSMC, it is necessary to reach $\delta_d = \frac{d\delta_d}{dt} = 0$ and $\delta_q = \frac{d\delta_q}{dt} = 0$. In the second method, the control inputs (U_d and U_q) can be built as shown in (7.15) and (7.16) [18]:

$$\begin{aligned} U_d &= u_{d1} + u_{d2} \\ u_{1d} &= -\beta\sqrt{|\delta_d|}\ \ sgn(\delta_d) \\ \dot{u}_{2d} &= -\alpha \mathrm{sgn}(\delta_d) \end{aligned} \tag{7.15}$$

and

$$\begin{aligned} U_q &= u_{q1} + u_{q2} \\ u_{q1} &= -\beta\sqrt{|\delta_q|}\ \ \mathrm{sgn}(\delta_q) \\ \dot{u}_{q2} &= -\alpha \mathrm{sgn}(\delta_q) \end{aligned} \tag{7.16}$$

It is obvious that each control input has two terms. The first one is the continuous-time function of the sliding surface, while the second one is the integral of the sliding surface. As a result, STSMC control is also continuous. Furthermore, the STSMC method does not require the derivatives of δ_d and δ_q. From (7.15) and (7.16), it is clear that the control inputs are independent of the system parameters. This feature makes the STSMC robust against disturbances. for stability issues, the controller gains must be strictly positives ($\beta > 0$ and $\alpha > 0$). The convergence to the sliding surface is achieved if β and α are selected as [18]:

$$\alpha > \frac{F_M}{H_L},\ \beta^2 \geq \frac{4F_M H_u(\alpha + F_M)}{H_L^3(\alpha - F_M)} \tag{7.17}$$

where F_M denotes the positive bound of A that must satisfy $F_M \geq |A|$. H_u and H_L are the upper and lower positive bounds of b that must satisfy $H_L \leq b \leq H_u$. A and b are functions of δ's second derivative, as defined by (7.18):

$$\frac{d^2\delta}{dt^2} = A + b \quad \frac{dU}{dt} \tag{7.18}$$

where A and b are given by (7.19):

$$A = \frac{-1}{L}\begin{bmatrix} \dfrac{dV_{g,d}}{dt} + R\dfrac{di_d}{dt} - LW\dfrac{di_q}{dt} \\ \dfrac{dV_{g,d}}{dt} + R\dfrac{di_d}{dt} + LW\dfrac{di_d}{dt} \end{bmatrix} \quad b = \frac{V_{in}}{2L} \tag{7.19}$$

7.4.2 Stability of the super twisting SMC

Now, the following Lyapunov function is defined by (7.20):

$$S = \frac{1}{2}\left(\delta_d^2 + \delta_q^2\right) \tag{7.20}$$

The derivative of (7.20) is given by (7.21):

$$\dot{S} = \delta_d \dot{\delta}_d + \delta_q \dot{\delta}_q \tag{7.21}$$

Replacing (7.21) by (7.12) and then by (7.15) and (7.16), we obtain (7.22):

$$\dot{S} = S_1 - \frac{V_{in}}{2L}\left(\beta\, \delta_d \sqrt{|\delta_d|}\, \mathrm{sgn}(\delta_d) + \alpha \int_0^t \delta_d\, \mathrm{sgn}(\delta_d)\, dt\right) + S_2$$
$$- \frac{V_{in}}{2L}\left(\beta\, \delta_q \sqrt{|\delta_q|}\, \mathrm{sgn}(\delta_q) + \alpha \int_0^t \delta_q\, \mathrm{sgn}(\delta_q)\, dt\right) \tag{7.22}$$

where S_1 and S_2 are given by (7.23):

$$S_1 = \frac{-\delta_d}{L}\left(V_{g,d} + R\, i_d + LW\, i_q\right), \quad S_2 = \frac{-\delta_q}{L}\left(V_{g,q} + R\, i_q + LW\, i_d\right) \tag{7.23}$$

Independently of S_1 and S_2 signs, as $\delta_d \mathrm{sgn}(\delta_d) > 0$, $\delta_q \mathrm{sgn}(\delta_q) > 0$, $\beta > 0$, and $\alpha > 0$, then choosing high values of β and α would result in $\dot{S} < 0$. The minimum values of α and β are set by using (7.17). In order to achieve optimum performances, these values could be tuned. By means of α and β gains, it is possible to adjust chattering, robustness, dynamic response, and steady-state error. While β is particularly effective at adjusting the speed of dynamic behaviour, α is vital to cancel the steady-state error. The chattering effect can be minimized since the STSMC generates a continuous control strategy instead of a discontinuous one [21].

The (direct and quadrature) *dq*-component of the control input in (7.15) and (7.16) clearly shows that while β is multiplied with $\sqrt{|\delta_d|}\, \mathrm{sgn}(\delta_d)$, α is multiplied with $\int \mathrm{sgn}(\delta_d)$. Consequently, the chattering which is caused by the discontinuities in δ_d and δ_q would be reduced by adjusting β rather than α. As shown by the control input in (7.15) and (7.16), the STSMC method is independent of the inductance value (L). After the elaboration of the control inputs in (7.15) and (7.16), PWM signals, then, are obtained by comparing the control inputs with a high-frequency triangular carrier signal. This leads to a fixed switching frequency.

7.5 Results and discussion

After the theoretical study illustrated above, a simulation under the MATLAB®/Simulink® environment is carried out in order to show the advantages of the proposed SMC. It should be noted that the results obtained in this study are found using the parameters shown in Table 7.2. The reference current i_d^{ref} is set to be 250 A (peak) while i_q^{ref} is set to 0. Moreover, Figure 7.2 depicts a detailed schematic diagram of the proposed STSMC method based on (7.15) and (7.16).

Figure 7.3 illustrates the steady-state voltage and current of phase A before the LCL filter stage. It is clear that the voltage at the output of the NPC inverter has three levels as well as a sinusoidal form of the current. A zoom on the shape of the current is illustrated in Figure 7.4. In addition, Figure 7.5 represents the FFT analysis of the current. The current in this case study has a THD = 1.1%. Furthermore,

Table 7.2 Simulation parameters in MATLAB®/Simulink®

Grid amplitude (Vrms Ph–Ph)	**415 V**
Frequency	50 Hz
V_{in}: inverter input voltage (DC)	800 V
Capacitors C1 and C2	3,300 e−4 F
Filter inductances (L1, L2)	45e−6 H
Filter capacitor (C)	80e−6 F
SMC gains (α, β)	1,000, 1,570,000
Sampling period	1 µs
Triangular carrier signal frequency	10 kHz

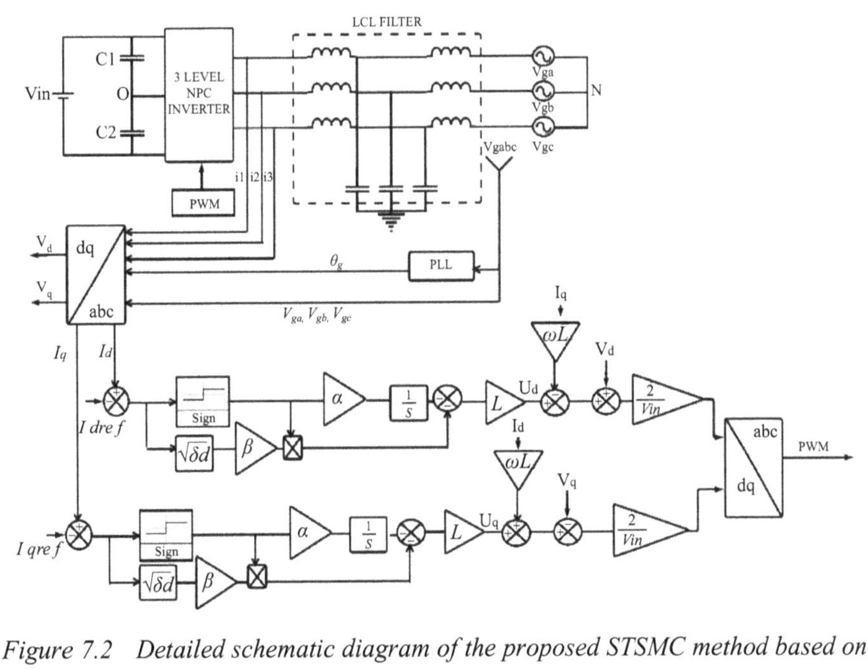

Figure 7.2 Detailed schematic diagram of the proposed STSMC method based on (7.15) and (7.16)

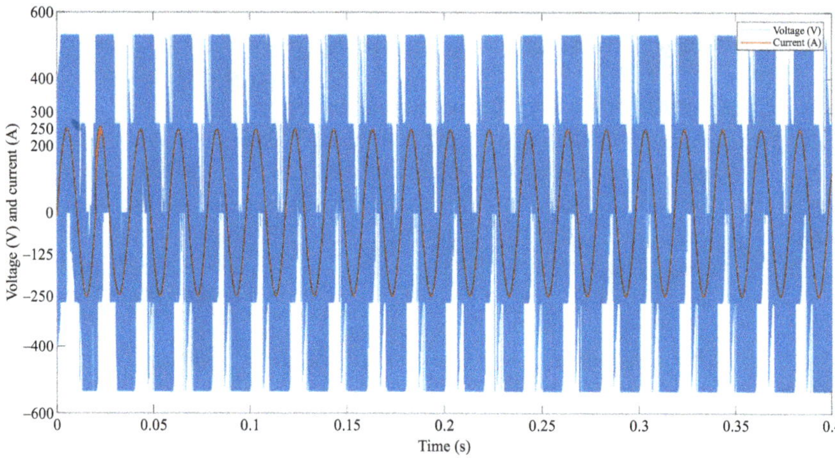

Figure 7.3 Steady-state phase A voltage and current before the LCL filter stage

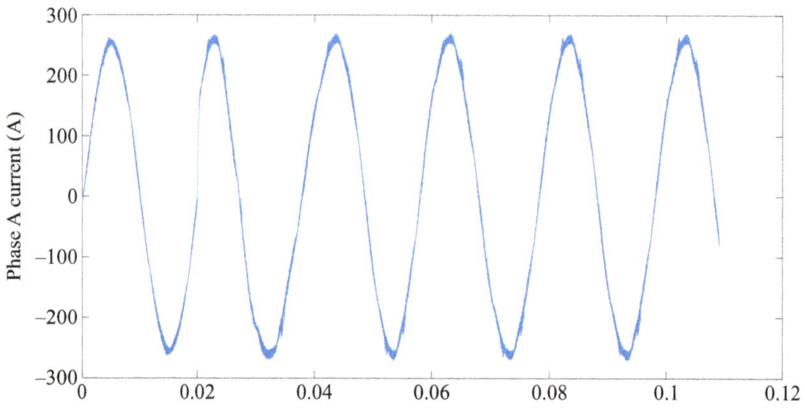

Figure 7.4 Zoom in phase A current

it is clear that there is no phase shift between current and voltage. As depicted in Figure 7.6, the voltage before the filter stage has a high THD having a value of 57.52%.

In Figure 7.7, the shape of the voltage and current of phase A after the LCL filter stage is represented. From this figure, it is clear that the voltage and the current are in phase. Moreover, the current follows properly the reference current (250 A peak, 50 Hz).

The THD of the current at the point of common connection to the electrical grid represents a rate of 0.53% as shown in Figure 7.8. This value is in good

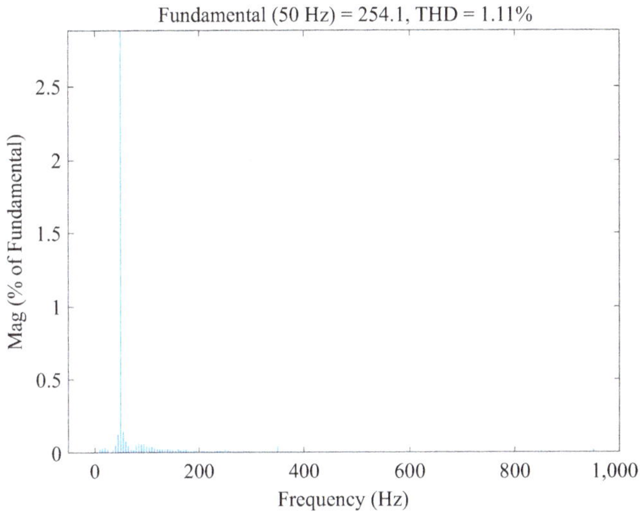

Figure 7.5 Current FFT analysis before the LCL filter stage

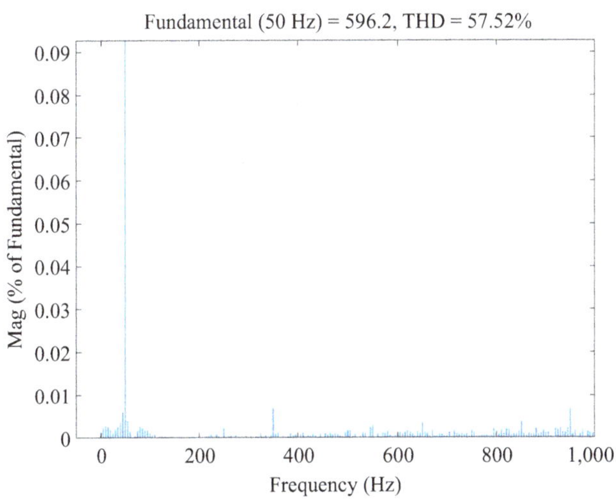

Figure 7.6 Phase A voltage FFT analysis before the LCL filter stage

agreement with interconnection specifications as mentioned above in the introduction section.

Furthermore, in order to assess the designed controller's fast tracking response and the dynamic performance, i_d^{ref} current is changed suddenly in different moments according to Figure 7.9. As shown in Figure 7.10, it is clear that the controller can follow correctly the reference current.

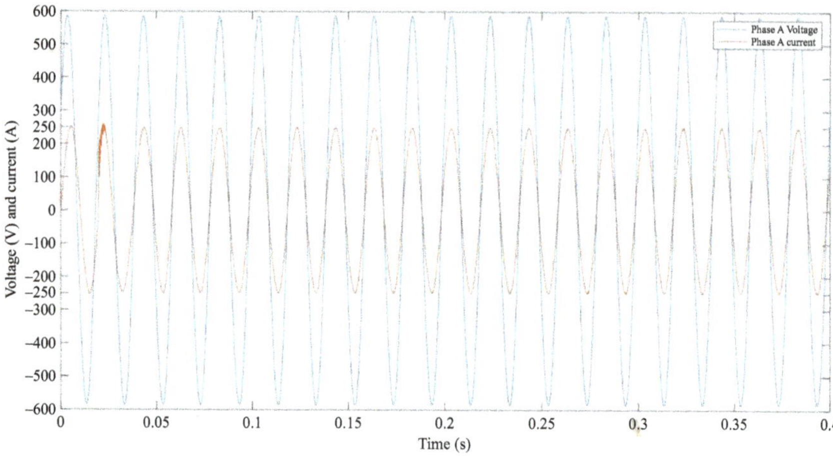

Figure 7.7 Steady-state voltage and current of phase A after the LCL filter stage

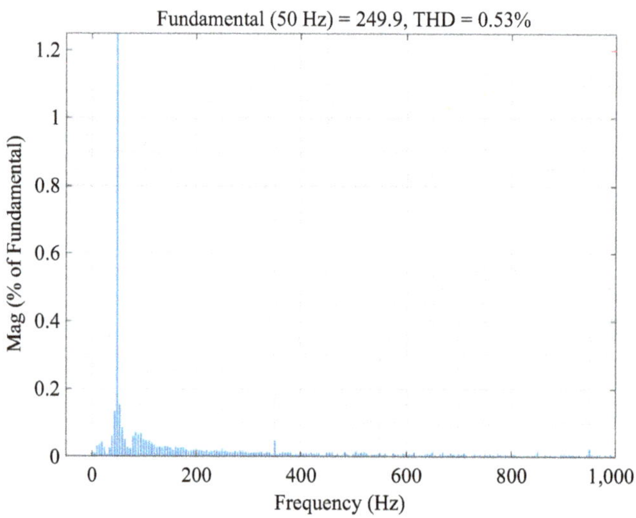

Figure 7.8 Current FFT analysis after the LCL filter stage

In Table 7.3, we vary the gain in order to see its effect on the dynamic response of the system. According to this table, β has a significant impact on the static error. Furthermore, as β increases, THD decreases. The value that generates near zero static error is for β = 1,570,000 regardless of the value of α.

In Figure 7.11, a sudden change in grid frequency from 50 Hz to 60 Hz is caused at 0.2 s–0.4 s. As shown in Figure 7.11, the injected current into the grid follows exactly the grid frequency. However, we can clearly see a small shift from 0.2 s up to 0.025 s. This is certainly due to the PLL which requires a transient time

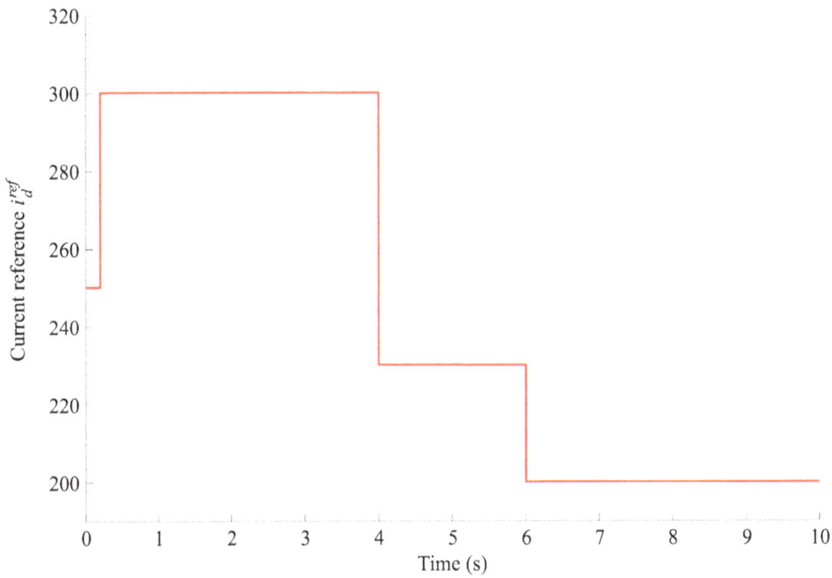

Figure 7.9 Current references i_d^{ref}

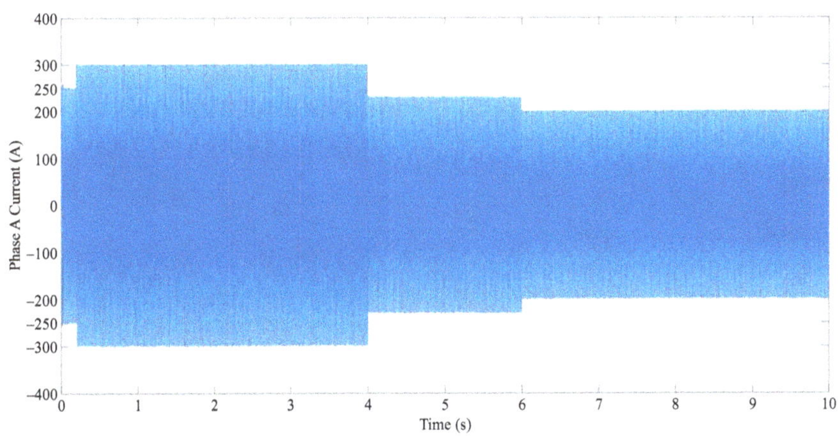

Figure 7.10 Steady-state phase A current for different i_d^{ref} current references

Table 7.3 Evolution of THD and current fundamental with respect to β

β	3,000,000	2,000,000	1,570,000	100,000
THD	0.46%	0.48%	0.53%	0.67%
Current fundamental (peak)	245.7 A	248.1 A	249.9 A	255.9 A

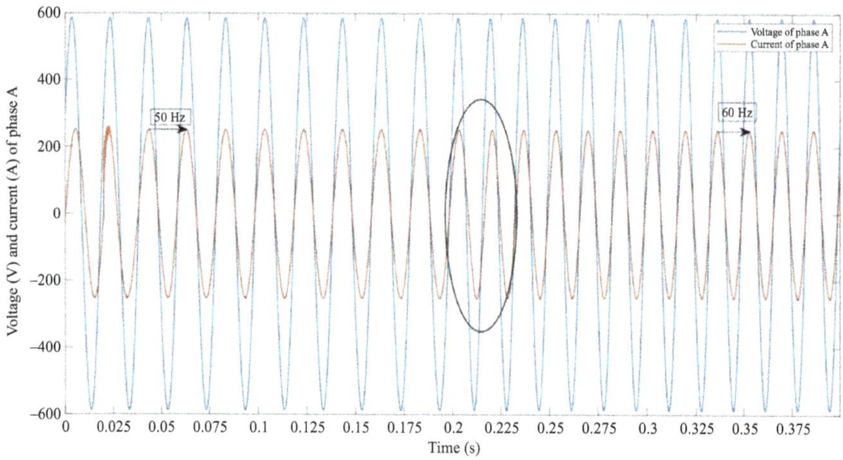

Figure 7.11 Steady-state voltage and current of phase A after a sudden change of frequency at 0.2 s

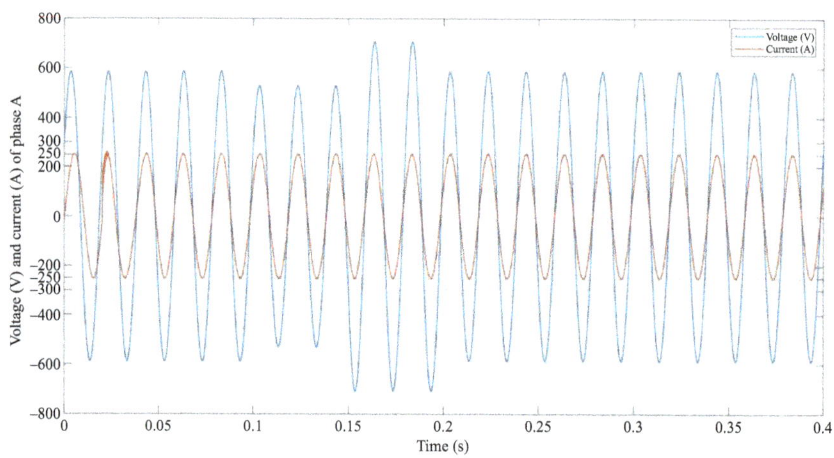

Figure 7.12 Steady-state voltage and current of phase A after a sudden change of the amplitude at 0.1s and 0.2 s

to extract the new parameters from the electrical grid. Then, we can see that there is an ideal synchronization between the voltage wave and the current wave.

In Figure 7.12, a sudden change in grid amplitude occurs at time 0.1 s from 586 V peak to 530 V in phase A only while the two other phases remain intact. At time 0.15 s, another change in amplitude is caused, this time from 586 V peak to 707.5 V. Then, the voltage returned to its initial value of 586 V at time 0.2 s. It is clear that the voltage and the current are always in phase and the amplitude of the injected current effectively follows the setpoint of 250 A peak. This shows the robustness of the STSMC used in this case.

7.6 Conclusion

In this chapter, a STSMC is designed to control a three-phase NPC inverter with an LCL filter. The control design ensures a regulated output current with a low THD value (0.51%). In fact, the use of Gao's reaching law is used to avoid chattering which is the main drawback of conventional SMC. The waveforms of the injected current are in phase with the voltage at the common point of connection even after sudden changes in the grid frequency and phase A voltage of the grid. This means that the STSMC has the ability to reject disturbances and uncertainties, such as load variations and grid disturbances, while maintaining a high level of performance. Moreover, the STSMC used in this study works with a fixed switching frequency.

References

[1] Zeb K, Uddin W, Khan MA, *et al.* A comprehensive review on inverter topologies and control strategies for grid connected photovoltaic system. *Renewable and Sustainable Energy Reviews* 2018;94:1120–41. https://doi.org/10.1016/j.rser.2018.06.053.

[2] Mahlooji MH, Mohammadi HR, and Rahimi M. A review on modeling and control of grid-connected photovoltaic inverters with LCL filter. *Renewable and Sustainable Energy Reviews* 2018;81:563–78. https://doi.org/10.1016/j.rser.2017.08.002.

[3] Ded A V, Maltsev VN, and Sikorski SP. Comparative analysis of the specifications on the power quality of the European Union and the Russian Federation. *Journal of Physics: Conference Series* 2018;998:1–6. https://doi.org/10.1088/1742-6596/998/1/012007.

[4] International Standard/IEC-61727IEC. Central Office GENEVA S. Photovoltaic (PV) Systems – Characteristics of the Utility Interface 2004.

[5] Barzegar-Kalashani M, Tousi B, Mahmud MA, and Farhadi-Kangarlu M. Higher-order sliding mode current controller for grid-connected distributed energy resources with LCL filters under unknown grid voltage conditions. *IET Generation, Transmission and Distribution* 2022;16:1592–606. https://doi.org/10.1049/gtd2.12384.

[6] Shin D, Kim HJ, Lee JP, Kim TJ, and Yoo DW. Design of LCL filter for improving robustness of grid-connected voltage source inverter. In: *2014 IEEE Energy Conversion Congress and Exposition, ECCE 2014*, 2014. p. 2940–6. https://doi.org/10.1109/ECCE.2014.6953798.

[7] Guzman R, De Vicuna LG, Camacho A, Matas J, Castilla M, and Miret J. Active damping control for a three phase grid-connected inverter using sliding mode control. In: *IECON Proceedings (Industrial Electronics Conference)*, 2013. p. 382–7. https://doi.org/10.1109/IECON.2013.6699166.

[8] Zeng Z, Yang H, Zhao R, and Cheng C. Topologies and control strategies of multi-functional grid-connected inverters for power quality enhancement: a comprehensive review. *Renewable and Sustainable Energy Reviews* 2013;24:223–70. https://doi.org/10.1016/j.rser.2013.03.033.

[9] Cortes P, Kazmierkowski MP, Kennel RM, Quevedo DE, and Rodriguez J. Predictive control in power electronics and drives. *IEEE Transactions on Industrial Electronics* 2008;55:4312–24. https://doi.org/10.1109/TIE.2008.2007480.

[10] Hong W, Tao G, and Wang H. Adaptive control techniques for three-phase grid-connected photovoltaic inverters. In: Precup R-E, Kamal T, and Zulqadar Hassan S (eds.), *Solar Photovoltaic Power Plants: Advanced Control and Optimization Techniques*, Singapore: Springer Singapore; 2019, p. 1–24. https://doi.org/10.1007/978-981-13-6151-7_1.

[11] Chowdhury VR and Kimball JW. Adaptive control of a three-phase grid connected inverter with near deadbeat response. In: *2021 IEEE Applied Power Electronics Conference and Exposition (APEC)*, 2021, p. 2698–701. https://doi.org/10.1109/APEC42165.2021.9486983.

[12] Espi JM, Castello J, García-Gil R, Garcera G, and Figueres E. An adaptive robust predictive current control for three-phase grid-connected inverters. *IEEE Transactions on Industrial Electronics* 2011;58:3537–46. https://doi.org/10.1109/TIE.2010.2089945.

[13] Ali Khan MY, Liu H, Yang Z, and Yuan X. A comprehensive review on grid connected photovoltaic inverters, their modulation techniques, and control strategies. *Energies* 2020;13:1–40. https://doi.org/10.3390/en13164185.

[14] Bendaoud M. Sliding mode control of boost rectifiers operating in discontinuous conduction mode for small wind power generators. *Wind Engineering* 2022;46:938–48. https://doi.org/10.1177/0309524X211060551.

[15] Fridman L, Barbot J-P, Plestan F (eds.), *Recent Trends in Sliding Mode Control*. Institution of Engineering and Technology; 2016. https://doi.org/10.1049/PBCE102E.

[16] Norsahperi NMH and Danapalasingam KA. An improved optimal integral sliding mode control for uncertain robotic manipulators with reduced tracking error, chattering, and energy consumption. *Mechanical Systems and Signal Processing* 2020;142:106747. https://doi.org/https://doi.org/10.1016/j.ymssp.2020.106747.

[17] Venkateswaran R, Yesudhas AA, Lee SR, and Joo YH. Integral sliding mode control for extracting stable output power and regulating DC-link voltage in PMVG-based wind turbine system. *International Journal of Electrical Power & Energy Systems* 2023;144:108482. https://doi.org/https://doi.org/10.1016/j.ijepes.2022.108482.

[18] Gao W, Wang Y, and Homaifa A. Discrete-time variable structure control systems. *IEEE Transactions on Industrial Electronics* 1995;42:117–22. https://doi.org/10.1109/41.370376.

[19] Sebaaly F, Vahedi H, Kanaan HY, Moubayed N, and Al-Haddad K. Sliding mode fixed frequency current controller design for grid-connected NPC

inverter. *IEEE Journal of Emerging and Selected Topics in Power Electronics* 2016;4:1397–405. https://doi.org/10.1109/JESTPE.2016.2586378.
[20] Pati AK and Sahoo NC. Adaptive super-twisting sliding mode control for a three-phase single-stage grid-connected differential boost inverter based photovoltaic system. *ISA Transactions* 2017;69:296–306. https://doi.org/https://doi.org/10.1016/j.isatra.2017.05.002.
[21] Chalanga A, Kamal S, Fridman LM, Bandyopadhyay B, and Moreno JA. Implementation of super-twisting control: super-twisting and higher order sliding-mode observer-based approaches. *IEEE Transactions on Industrial Electronics* 2016;63:3677–85. https://doi.org/10.1109/TIE.2016.2523913.
[22] Sharma B, Dahiya R, and Nakka J. Effective grid connected power injection scheme using multilevel inverter based hybrid wind solar energy conversion system. *Electric Power Systems Research* 2019;171:1–14. https://doi.org/https://doi.org/10.1016/j.epsr.2019.01.044.
[23] Fernão Pires V, Cordeiro A, Foito D, and Fernando Silva J. Three-phase multilevel inverter for grid-connected distributed photovoltaic systems based in three three-phase two-level inverters. *Solar Energy* 2018;174:1026–34. https://doi.org/https://doi.org/10.1016/j.solener.2018.09.083.
[24] Arulkumar K, Vijayakumar D, and Palanisamy K. Modeling and control strategy of three phase neutral point clamped multilevel PV inverter connected to the grid. *Journal of Building Engineering* 2015;3:195–202. https://doi.org/https://doi.org/10.1016/j.jobe.2015.06.001.
[25] Laamiri S, Ghanes M, and Santomenna G. Observer based direct control strategy for a multi-level three phase flying-capacitor inverter. *Control Engineering Practice* 2019;86:155–65. https://doi.org/https://doi.org/10.1016/j.conengprac.2019.03.011.
[26] Katir H, Abouloifa A, Noussi K, *et al.* PV-powered grid-tied fault tolerant double-stage DC-AC conversion chain based on a cascaded H-bridge multilevel inverter. *IFAC-Papers* 2022;55:103–8. https://doi.org/https://doi.org/10.1016/j.ifacol.2022.07.295.
[27] Wu W, Wang F, and Wang Y. A novel efficient T type three level neutral-point-clamped inverter for renewable energy system. In *2014 International Power Electronics Conference (IPEC-Hiroshima 2014 – ECCE ASIA)*, 2014, p. 470–4. https://doi.org/10.1109/IPEC.2014.6869625.
[28] Baker RH. Bridge Converter Circuit 1981. https://patents.google.com/patent/US4270163A/en.
[29] Nabae A, Takahashi I, and Akagi H. A new neutral-point-clamped PWM inverter. *IEEE Transactions on Industry Applications* 1981;IA-17:518–23. https://doi.org/10.1109/TIA.1981.4503992.

Chapter 8

Neuro control of grid-connected three-phase inverters

Muhammad Maaruf[1]

Renewable power is transferred to the utility grid via three-phase inverters. The transfer of the desired power from the inverters with high dynamic performance requires an advanced control strategy that can suppress unknown time-varying disturbances and ensure accurate tracking performance. This chapter presents a neuro-sliding mode control strategy to achieve this objective. A radial basis function neural network (RBFNN) architecture is utilized to estimate the unknown time-varying disturbances. A Lyapunov candidate function is used to highlight the asymptotic convergence of the tracking errors. Simulation results demonstrate the impact of the suggested control method.

8.1 Introduction

Three-phase inverters connected to the grid are performing various roles such as transferring generated renewable power from renewable energy sources to utility grids and several AC loads, and interfacing several microgrids with the main grid [1,2]. To execute these tasks according to the requirements set by the national standards for grid-connected inverters, effective control and modulation algorithms are required [3]. The preferred modulation method is the space vector pulse width modulation (SVPWM) due to its less complex design, low harmonic proportion, and fixed switching frequency [4]. The fundamental operation of the grid-connected inverters requires synchronization of the inverter output parameters with the grid, which is performed by the phase-locked loop (PLL) [5]. Since the phase angle of a PLL can be measured, it is used for current and voltage transformation in the d–q synchronous reference frame. The active and reactive power injected into the grid is regulated by controlling the d–q current of the inverter.

Several control approaches have been proposed for the three-phase inverters to facilitate the transfer of the required active and reactive power with acceptable dynamic performances [6]. Proportional–integral (PI) and proportional–integral–

[1]Control and Instrumentation Engineering Department & Center for Smart Mobility and Logistics, King Fahd University of Petroleum & Minerals, Saudi Arabia

derivative (PID) control strategies are commonly applied to the grid-tied inverters [7–10]. However, PI and PID techniques are mainly employed for the linear model of the grid-tied inverters and it is difficult to obtain some performances such as short transient period, low current harmonic distortion, and fast convergence to the desired state in the presence of external disturbances and dynamic uncertainties. Another popular control technique applied to the three-phase inverters is model predictive control (MPC) [11–15]. MPC is an optimal control strategy that minimizes a cost function to determine the control signals sent to the inverter using a predicted model. However, most of the MPC implemented for power injection into the grid are linear and require the exact system model. Various nonlinear control techniques such as sliding mode control (SMC) [16–28], passivity-based control [29–32], synergetic control [33,34], backstepping control [35,36], and feedback linearization control [37,38] have been applied to the gird-tied inverters to enhance their performance.

Among the aforesaid nonlinear control methods, the SMC has gained more attention because of its simplicity, fast response, and robustness to disturbances. The authors in [16–19,21,22,24–26] assumed that the disturbances in the grid-tied inverters are constants and their upper-bounds are known while the authors in [20,23,27,28] used adaptation laws to estimate the upper-bounds of the disturbances. However, the adaptive laws are only effective in estimating constant disturbance. In practical applications, the disturbances in the grid-tied inverter systems are time varying because of the fluctuations in renewable power and consumption.

Motivated by the above discussion, in this chapter, the control of the current injected into the grid from the three-phase inverter is presented. The desired currents of the d- and q-axis reference frames are derived from the required active and reactive power injected into the grid, respectively. An SMC technique is used to achieve the control aims and address the unknown dynamic uncertainties and time varying disturbances in the system. Due to the universal approximation property of the RBFNN, it is used to estimate the uncertain functions together with the unknown time-varying disturbances. Simulation results are given to demonstrate the efficacy of the designed controller.

The remainder of the chapter is arranged as follows: the description of the system and its dynamic modeling are provided in Section 8.2. The proposed neuro controller is developed for the grid-tied three-phase inverters in Section 8.3. The discussion of the simulation results is provided in Section 8.4. In Section 8.5, the conclusion of this chapter is highlighted.

8.2 System description

Figure 8.1 depicts the topology of the grid-tied three-phase inverter and its equivalent single-phase circuit. The inverter is fed by a DC voltage source (V_{dc}) while its output three-phase voltage $(V_{conv\{abc\}})$ and current $(I_{conv\{abc\}})$ are filtered by an L-filter with inductance L_g and resistance R_g. The filter lies between the AC side of the inverter and the grid circuit with three-phase voltage $(V_{g\{abc\}})$ and

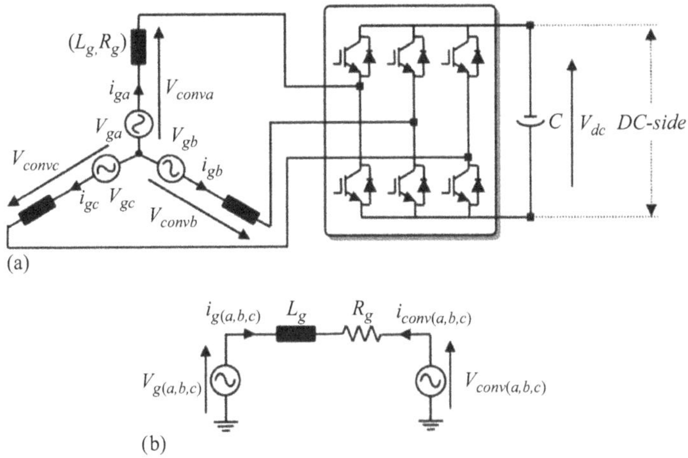

Figure 8.1 Schematic diagram of a three-phase grid-tied inverter (a) together with the equivalent single-phase circuit (b)

current $(I_{g\{abc\}})$. The inverter output is regulated with space vector modulation (SVM). The mathematical model of the grid-connected inverter can be derived by transforming the *abc* variable model into *dq* synchronous reference frame as follows [1,39–41]:

$$L_g \frac{di_{gd}}{dt} = \frac{1}{L_g} V_{gd} - R_g i_{gd} + \omega_g L_g i_{gq} - V_{convd} + \delta_d \tag{8.1}$$

$$L_g \frac{di_{gq}}{dt} = V_{gq} - R_g i_{gq} - \omega_g L_g i_{gd} - V_{convq} + \delta_q \tag{8.2}$$

where δ_d and δ_q are time-varying disturbances. The inverter synchronizes the utility grid with the aid of a PLL. The PLL achieves the synchronization by forcing the phase (θ) of the grid voltage to zero. The dynamic equation of the grid voltage phase is given by [42]:

$$\dot{\theta} = \omega_g \tag{8.3}$$

where ω_g is the grid angular frequency. It is obvious that (8.1) and (8.2) are nonlinear equations. The reference angle for θ is set as $\theta^r = 0$ and the tracking error is given calculated as:

$$\zeta_\theta = \theta^r - \theta \tag{8.4}$$

The controller that will ensure θ converges to zero is designed as follows:

$$\omega_g = \dot{\theta}^r + \left(K_p + \frac{k_I}{s}\right)\zeta_\theta \tag{8.5}$$

The active and reactive power equations of the system can be established as follows:

$$P = 1.5[V_{gd}i_{gd} + V_{gq}i_{gq}] \tag{8.6}$$

$$Q = 1.5[V_{gq}i_{gd} - V_{gd}i_{gq}] \tag{8.7}$$

By aligning the *d*-axis with the grid voltage vector, (8.6) and (8.7) become:

$$P = 1.5V_{gd}i_{gd} \tag{8.8}$$

$$Q = -1.5V_{gd}i_{gq} \tag{8.9}$$

8.3 Control design

In this section, a neuro-adaptive sliding mode control of the current transferred to the grid is designed. The control block diagram of the closed-loop system is shown in Figure 8.2. It is desired that the maximum active power and the minimum reactive power ($Q^r = 0$) are transferred to the grid circuit. The reference currents for i_{gd} and i_{gq} can be computed from (8.8) and (8.9), respectively, as follows:

$$i_{gd}^r = \frac{P^r}{V_{gd}} \tag{8.10}$$

$$i_{gq}^r = \frac{Q^r}{V_{gd}} \tag{8.11}$$

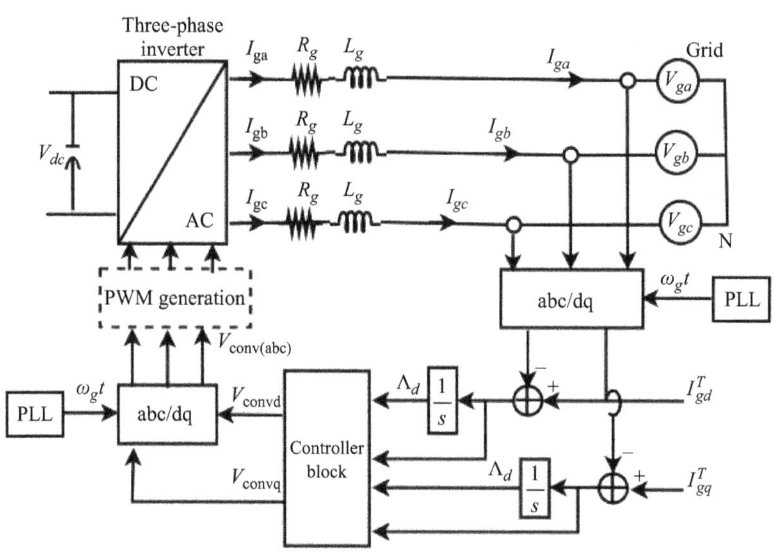

Figure 8.2 The control block diagram of the three-phase grid-connected inverter

8.3.1 Neural network approximation

RBFNNs are widely used to approximate continuous time and smooth nonlinear functions due to their universal approximation property and simplicity. A continuous and smooth function $J(\chi)$ can be approximated with an RBFNN as follows:

$$J(\chi) = W^T\beta(\chi) + \varepsilon(X) \tag{8.12}$$

where $W = \left[w_1\ w_2 \ldots w_n\right]^T \in \mathbb{R}^n$, n is the number of hidden nodes, $\beta(\chi)$ is the basis function of the hidden nodes, and ε is the approximation error. The Gaussian basis function vector is

$$\beta(\chi) = exp\left(-\frac{(\chi-\eta)(\chi-\eta)}{b^2}\right), \quad i= 1,2,\ldots, m \tag{8.13}$$

where b is the width of the Gaussian function and η is the center of the receptive field.

8.3.2 Neuro sliding mode control design

The sliding mode surfaces for i_{gd} and i_{gq} are, respectively, designed as follows:

$$\zeta_d = \Lambda_d \int (i_{gd}^r - i_{gd})dt + (i_{gd}^r - i_{gd}) \tag{8.14}$$

$$\zeta_q = \Lambda_q \int (i_{gq}^r - i_{gq})dt + (i_{gq}^r - i_{gq}) \tag{8.15}$$

where Λ_d and Λ_q are positive constants. The derivatives of ζ_d and ζ_q yield:

$$\dot{\zeta}_d = \Lambda_d(i_{gd}^r - i_{gd}) + \frac{d}{dt}(i_{gd}^r - i_{gd}) = \Lambda_d(i_{gd}^r - i_{gd}) + \frac{di_{gd}^r}{dt} - F_d + \frac{1}{L_g}V_{convd} \tag{8.16}$$

$$\dot{\zeta}_q = \Lambda_q(i_{gq}^r - i_{gq}) + \frac{d}{dt}(i_{gq}^r - i_{gq}) = \Lambda_q(i_{gq}^r - i_{gq}) + \frac{di_{gq}^r}{dt} - F_q + \frac{1}{L_g}V_{convq} \tag{8.17}$$

where

$$F_d = \frac{1}{L_g}\left(V_{gd} - R_g i_{gd} + \omega_g L_g i_{gq} + \delta_d\right) \tag{8.18}$$

$$F_q = \frac{1}{L_g}\left(V_{gq} - R_g i_{gq} - \omega_g L_g i_{gd} + \delta_q\right) \tag{8.19}$$

The uncertain nonlinear functions can be approximated by RBFNN as follows:

$$F_d = W_d^T\beta(\chi_d) + \varepsilon_d \tag{8.20}$$

$$F_q = W_q^T \beta(\chi_q) + \varepsilon_q \tag{8.21}$$

The neural network estimates of the nonlinear functions are thus:

$$\widehat{F}_d = \widehat{W}_d^T \beta(\chi_d) \tag{8.22}$$

$$\widehat{F}_q = \widehat{W}_q^T \beta(\chi_q) \tag{8.23}$$

The wavelet neural network estimation errors are given as:

$$\widetilde{F}_d = F_d - \widehat{F}_d = W_d^T \beta(\chi_d) + \varepsilon_d - \widehat{W}_d^T \beta(\chi_d) = \widetilde{W}_d^T \beta(\chi_d) + \varepsilon_d \tag{8.24}$$

$$\widetilde{F}_q = F_q - \widehat{F}_q = W_q^T \beta(\chi_q) + \varepsilon_q - \widehat{W}_q^T \beta(\chi_q) = \widetilde{W}_q^T \beta(\chi_q) + \varepsilon_q \tag{8.25}$$

The robust control laws are defined as follows:

$$\dot{\zeta}_d = -\sigma_d\ (\zeta_d) - k_d\ sign(\zeta_d) \tag{8.26}$$

$$\dot{\zeta}_q = -\sigma_q\ (\zeta_q) - k_q\ sign(\zeta_q) \tag{8.27}$$

The control inputs for i_{gd} and i_{gq} are designed as follows:

$$V_{convd} = -L_d \left[\Lambda_d (i_{gd}^r - i_{gd}) + \frac{di_{gd}^r}{dt} - \widehat{F}_d + \sigma_d\ (\zeta_d) + k_d\ sign(\zeta_d) \right] \tag{8.28}$$

$$V_{convq} = -L_q \left[\Lambda_q (i_{gq}^r - i_{gq}) + \frac{di_{gq}^r}{dt} - \widehat{F}_q + \sigma_q\ (\zeta_q) + k_q\ sign(\zeta_q) \right] \tag{8.29}$$

The update laws for the weight vectors are designed as:

$$\dot{\widehat{W}}_d = -\gamma_d \zeta_d \beta(\chi_d) \tag{8.30}$$

$$\dot{\widehat{W}}_q = -\gamma_q \zeta_q \beta(\chi_q) \tag{8.31}$$

Inserting (8.28) and (8.29) into (8.16) and (8.17), respectively, gives:

$$\dot{\zeta}_d = -\sigma_d \zeta_d + \widetilde{W}_d^T \beta(\chi_d) + \varepsilon_d \tag{8.32}$$

$$\zeta_q = -\sigma_q \zeta_q + \widetilde{W}_q^T \beta(\chi_q) + \varepsilon_q \tag{8.33}$$

Consider the following candidate Lyapunov function:

$$T = \frac{1}{2}\zeta_d^2 + \frac{1}{2}\zeta_q^2 + \frac{1}{2\gamma_d}\widetilde{W}_d^2 + \frac{1}{2\gamma_q}\widetilde{W}_q^2 \tag{8.34}$$

Differentiating T with respect to time gives:

$$\dot{T} = \zeta_d\dot{\zeta}_d + \zeta_q\dot{\zeta}_q + \frac{1}{\gamma_d}\widetilde{W}_d^T\dot{\widetilde{W}}_d + \frac{1}{\gamma_q}\widetilde{W}_q^T\dot{\widetilde{W}}_q$$

$$= -\sigma_d\zeta_d^2 - \sigma_q\zeta_q^2 + \widetilde{W}_d^T\left(\zeta_d\beta(\chi_d) + \frac{\dot{\widehat{W}}_d}{\gamma_d}\right) + \widetilde{W}_q^T\left(\zeta_q\beta(\chi_q) + \frac{\dot{\widehat{W}}_q}{\gamma_q}\right) + \zeta_d\varepsilon_d$$

$$+ \zeta_q\varepsilon_q - k_d\,\zeta_d\,sign(\zeta_d) - k_q\,\zeta_q\,sign(\zeta_q) \tag{8.35}$$

Substituting the adaptive laws (8.30) and (8.30) into (8.35), one has:

$$\dot{T} \leq -\sigma_d\zeta_d^2 - \sigma_q\zeta_q^2 + |\zeta_d|(\varepsilon_d - k_d) + |\zeta_q|(\varepsilon_q - k_q) \tag{8.36}$$

Choosing $k_d \geq \varepsilon_d$ and $k_q \geq \varepsilon_q$, one has:

$$\dot{T} \leq -\sigma_d\zeta_d^2 - \sigma_q\zeta_q^2 \tag{8.37}$$

Therefore, the tracking errors in the closed-loop system are asymptotically stable.

8.4 Simulation results

This section discusses the results obtained by implementing the neuro controller to the grid-tied three-phase inverter in MATLAB®/Simulink® platform. The parameters of the system are given in Table 8.1. The gains of the controller are provided in Table 8.2.

Table 8.1 Parameters of the system

Parameters	**Values**
L_g	10^{-2} H
R_g	50×10^{-2} Ω
V_{gd}	311 V
V_{gq}	311 V
δ_d	$0.5 + 0.1\ \sin\ (2t) + e^{-2t}$
δ_q	$0.5 + 0.1\ \sin\ (2t) + e^{-2t}$

Table 8.2 Parameters of the neuro controller

Parameters	**Values**
λ_d, λ_q	25, 15
σ_d, σ_q	18, 11
k_d, k_q	2, 2
γ_d, γ_q	0.01, 0.01

As shown in Figure 8.3, the grid voltage is distorted by 8% of the 5th-order and 6% of the 7th-order harmonics for $0.25 \leq t \leq 0.35$ s. The reference active and the reactive power transferred to the grid from the three-phase inverter are presented in Figure 8.4. The reference active power increases from 1,200 W to 1,500 W at the time 0.15 s. It is clear that the distorted grid voltage introduces some ripples in the active and reactive power. However, the neuro controller is able to lessen the impact of the harmonic distortion and maintain the tracking of the desired injected power.

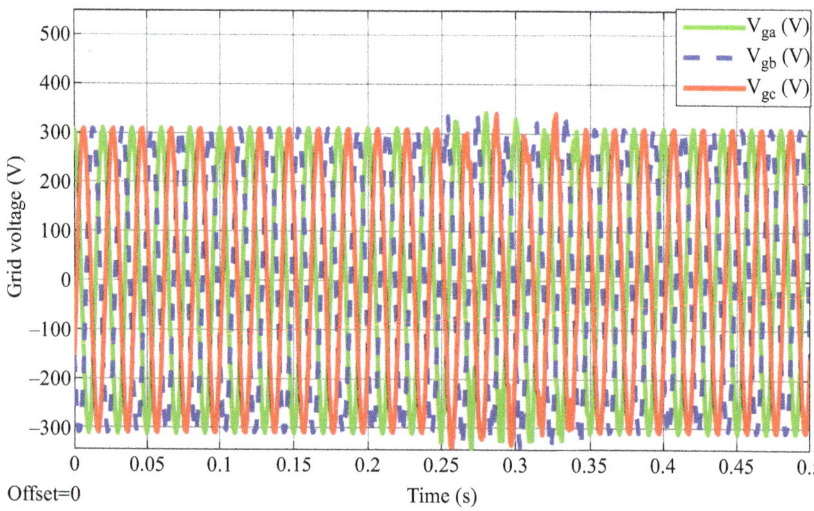

Figure 8.3 Evolution of the three-phase grid voltage with harmonic distortion

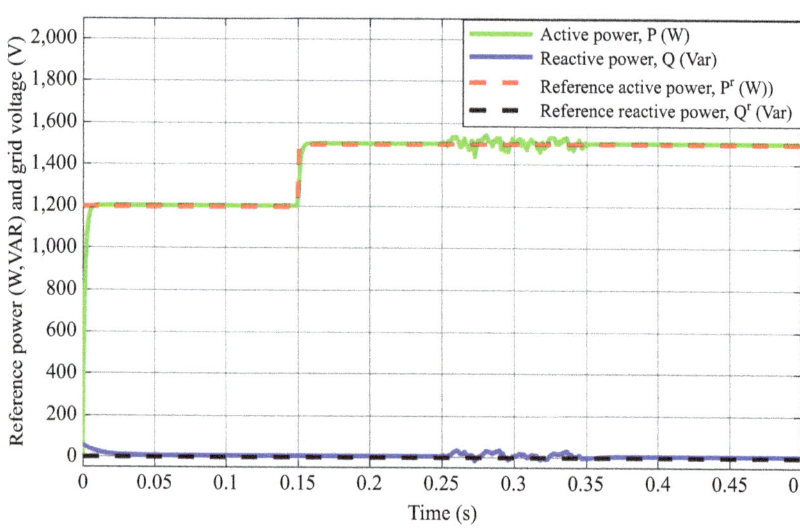

Figure 8.4 Tracking performance of the active and reactive power

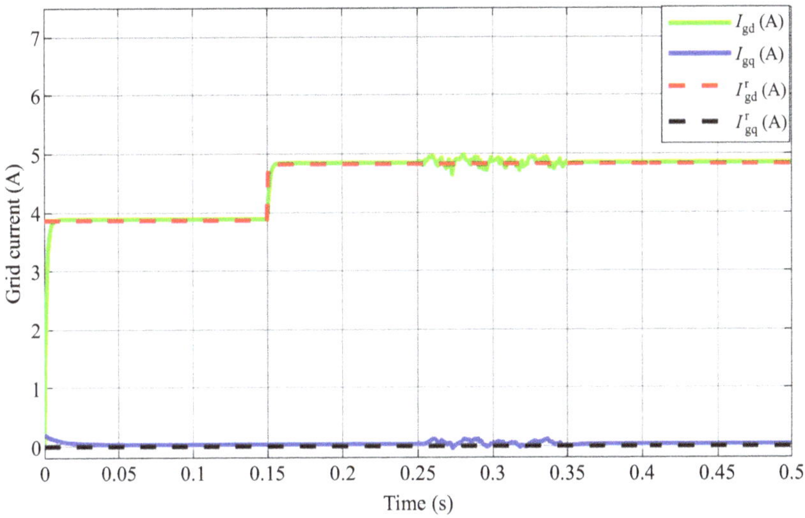

Figure 8.5 Tracking of the d- and q-axis components of the grid current

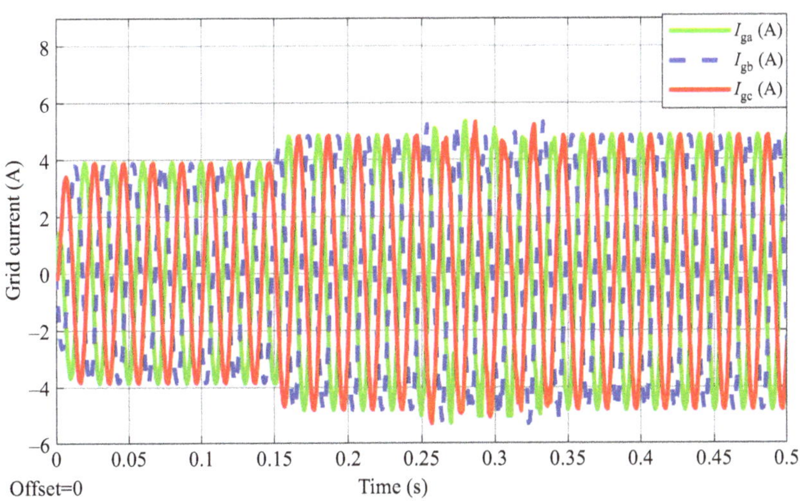

Figure 8.6 Evolution of the three-phase grid current with harmonic distortion

The reference *d*- and *q*-axis currents injected into the grid are calculated according to (8.10) and (8.11), respectively. Despite the presence of harmonic distortions and unknown time-varying disturbances, the neuro controller is able to achieve excellent regulation of the injected currents as shown in Figure 8.5. The severity of the harmonics distortion is mitigated and it can be seen that the *d*-axis current tracks the step-change of its reference whereas the *q*-axis current converge to zero. This will ensure the unity power factor operation of the grid. The evolution of the three-phase grid current is shown in Figure 8.6. It is clear that the amplitude of the three-phase current increases at the time of 0.15 s where the active power increases to 1,500 W.

8.5 Conclusion

This chapter investigated the sliding mode control of a three-phase inverter connected to the grid. This controller can ensure the asymptotic convergence of the injected currents to their respective reference values and tackle the dynamic uncertainties and the unknown time-varying disturbances in the system at all times. An RBFNN has been employed to approximate the dynamic uncertainties alongside the time-varying disturbances. The asymptotic stability of the system has been proven using a candidate Lyapunov function. The simulation results have shown that the controller can achieve accurate tracking of the current transferred to the grid irrespective of the unknown time-varying disturbances.

References

[1] Hassine IMB, Naouar MW, and Mrabet-Bellaaj N. Model predictive-sliding mode control for three-phase grid-connected converters. *IEEE Transactions on Industrial Electronics.* 2016;64(2):1341–1349.
[2] He Y, Guo S, Zhou J, *et al.* The many-objective optimal design of renewable energy cogeneration system. *Energy.* 2021;234:121244–68.
[3] Tang L, Han Y, Yang P, *et al.* A review of voltage sag control measures and equipment in power systems. *Energy Reports.* 2022;8:207–216.
[4] Chen M, Zhou D, Tayyebi A, *et al.* Generalized multivariable grid-forming control design for power converters. *IEEE Transactions on Smart Grid.* 2022;13(4):2873–2885.
[5] Guo L, Xu Z, Li Y, *et al.* An inductance online identification-based model predictive control method for grid-connected inverters with an improved phase-locked loop. *IEEE Transactions on Transportation Electrification.* 2021;8(2):2695–2709.
[6] Wu C, Xiong X, and Blaabjerg F. Impedance analysis of voltage source converter based on voltage modulated matrix. In: *2021 IEEE Energy Conversion Congress and Exposition (ECCE).* IEEE; 2021. p. 887–892.
[7] Gui Y, Wang X, Blaabjerg F, *et al.* Control of grid-connected voltage-source converters: the relationship between direct-power control and vector-current control. *IEEE Industrial Electronics Magazine.* 2019;13(2):31–40.
[8] Rey-Boué AB, Guerrero-Rodrguez NF, Stöckl J, *et al.* Modeling and design of the vector control for a three-phase single-stage grid-connected PV system with LVRT capability according to the Spanish grid code. *Energies.* 2019;12(15):2899–3017.

[9] Azzoug Y, Sahraoui M, Pusca R, *et al.* High-performance vector control without AC phase current sensors for induction motor drives: simulation and real-time implementation. *ISA Transactions.* 2021;109:295–306.
[10] Praiselin W and Edward JB. Voltage profile improvement of solar PV grid-connected inverter with micro grid operation using PI controller. *Energy Procedia.* 2017;117:104–111.
[11] Güler N and Irmak E. MPPT based model predictive control of grid connected inverter for PV systems. In: *2019 8th International Conference on Renewable Energy Research and Applications (ICRERA).* IEEE; 2019. p. 982–986.
[12] Golzari S, Rashidi F, and Farahani HF. A Lyapunov function based model predictive control for three phase grid connected photovoltaic converters. *Solar Energy.* 2019;181:222–233.
[13] Long B, Cao T, Fang W, *et al.* Model predictive control of a three-phase two-level four-leg grid-connected converter based on sphere decoding method. *IEEE Transactions on Power Electronics.* 2020;36(2):2283–2297.
[14] Dharmasena S, Olowu TO, and Sarwat AI. A low-complexity FS-MPDPC with extended voltage set for grid-connected converters. *IET Energy Systems Integration.* 2021;3(4):413–425.
[15] Andino J, Ayala P, Llanos-Proaño J, *et al.* Constrained modulated model predictive control for a three-phase three-level voltage source inverter. *IEEE Access.* 2022;10:10673–10687.
[16] Gui Y, Xu Q, Blaabjerg F, *et al.* Sliding mode control with grid voltage modulated DPC for voltage source inverters under distorted grid voltage. *CPSS Transactions on Power Electronics and Applications.* 2019;4(3):244–254.
[17] Dang C, Tong X, and Song W. Sliding-mode control in dq-frame for a three-phase grid-connected inverter with LCL-filter. *Journal of the Franklin Institute.* 2020;357(15):10159–10174.
[18] Benbouhenni H, Mehedi F, and Soufiane L. New direct power synergetic-SMC technique based PWM for DFIG integrated to a variable speed dual-rotor wind power. *Automatika.* 2022;63(4):718–731.
[19] Shen X, Liu J, Luo W, *et al.* High-performance second-order sliding mode control for NPC converters. *IEEE Transactions on Industrial Informatics.* 2020;16(8):5345–5356.
[20] Xu D, Wang G, Yan W, *et al.* A novel adaptive command-filtered back-stepping sliding mode control for PV grid-connected system with energy storage. *Solar Energy.* 2019;178:222–230.
[21] Huang X, Wang K, Fan B, *et al.* Robust current control of grid-tied inverters for renewable energy integration under non-ideal grid conditions. *IEEE Transactions on Sustainable Energy.* 2020;11(1):477–488.
[22] Ozdemir S, Altin N, Sefa I, *et al.* Super twisting sliding mode control of three-phase grid-tied neutral point clamped inverters. *ISA Transactions.* 2022;125:547–559.
[23] Arshad MH, El-Farik S, Abido M, *et al.* Genetic algorithm tuned adaptive discrete-time sliding mode controller for grid-connected inverter with an LCL filter. *Energy Reports.* 2022;8:623–640.

[24] Aillane A, Dahech K, Chouder A, *et al.* Control of a three-phase grid-connected inverter based on super-twisting sliding mode algorithm. In: *2022 19th International Multi-Conference on Systems, Signals & Devices (SSD). IEEE; 2022*. p. 1186–1192.

[25] Memije D, Carranza O, Rodríguez JJ, *et al.* Reduction of grid background harmonics in three-phase inverters by applying sliding mode control. In: *2020 IEEE 29th International Symposium on Industrial Electronics (ISIE)*. IEEE; 2020. p. 191–196.

[26] Altin N, Ozdemir S, Sefa I, *et al.* Second-order sliding mode control of three-phase three-level grid-connected neutral point clamped inverters. In: *2021 13th International Conference on Electronics, Computers and Artificial Intelligence (ECAI)*. IEEE; 2021. p. 1–6.

[27] Hollweg GV, de Oliveira Evald PJD, Mattos E, *et al.* Feasibility assessment of adaptive sliding mode controllers for grid-tied inverters with LCL filter. *Journal of Control, Automation and Electrical Systems.* 2022;33(2):434–447.

[28] Vieira Hollweg G, Dias de Oliveira Evald P, Varella Tambara R, *et al.* Adaptive super-twisting sliding mode for DC-AC converters in very weak grids. *International Journal of Electronics.* 2022;1–26.

[29] Abed ZM, Hassan TK, and Hameed KR. Analysis and design of photovoltaic three-phase grid-connected inverter using passivity-based control. *International Journal of Power Electronics and Drive Systems.* 2022;13(1):167.

[30] Li J, Zhang Y, Zhao Y, *et al.* An improved three-stages cascading passivity based control of grid-connected LCL converter in unbalanced weak grid condition. *IEEE Access.* 2021;9:89497–89506.

[31] Debnath R, Gupta G, and Kumar D. Lyapunov-Krasovskii passivity based stability analysis of grid-tied inverters. *International Journal of Electrical Power & Energy Systems.* 2022;143:108460.

[32] Özbay H, Öncü S, and Kesler M. SMC-DPC based active and reactive power control of grid-tied three phase inverter for PV systems. *International Journal of Hydrogen Energy.* 2017;42(28):17713–17722.

[33] Mehiri A, Bettayeb M, and Hamid AK. Fractional nonlinear synergetic control for three phase inverter tied to PV system. In: *2019 8th International Conference on Modeling Simulation and Applied Optimization (ICMSAO)*. IEEE; 2019. p. 1–5.

[34] Benbouhenni H and Bizon N. Terminal synergetic control for direct active and reactive powers in asynchronous generator-based dual-rotor wind power systems. *Electronics.* 2021;10(16):1880–1903.

[35] Ahmad A, Ullah N, Ahmed N, *et al.* Robust control of grid-tied parallel inverters using nonlinear backstepping approach. *IEEE Access.* 2018;7:111982–111992.

[36] Ullah N and Al Ahmadi AA. Variable structure back-stepping control of two-stage three phase grid connected PV inverter. *Periodica Polytechnica Electrical Engineering and Computer Science.* 2020;64(3):239–246.

[37] Chowdhury VR and Kimball JW. Grid voltage estimation and feedback linearization based control of a three phase grid connected inverter under

unbalanced grid conditions with LCL filter. In: *2019 IEEE Energy Conversion Congress and Exposition (ECCE)*. IEEE; 2019. p. 2979–2984.
[38] Yang L, Feng C, and Liu J. Control design of LCL type grid-connected inverter based on state feedback linearization. *Electronics*. 2019;8(8):877.
[39] Gui Y, Wang X, and Blaabjerg F. Vector current control derived from direct power control for grid-connected inverters. *IEEE Transactions on Power Electronics*. 2018;34(9):9224–9235.
[40] Mejía-Ruiz GE, Rodríguez J, Paternina M, *et al*. Grid-connected three-phase inverter system with LCL filter: model, control and experimental results. In: *2019 IEEE PES Innovative Smart Grid Technologies Conference-Latin America (ISGT Latin America)*. IEEE; 2019. p. 1–6.
[41] Younsi S and Hamrouni N. Control of grid connected three-phase inverter for hybrid renewable systems using sliding mode controller. *International Journal of Advanced Computer Science and Applications*. 2018;9(11):336–342.
[42] Radwan AAA and Mohamed YARI. Grid-connected wind-solar cogeneration using back-to-back voltage-source converters. *IEEE Transactions on Sustainable Energy*. 2020;11(1):315–325.

Chapter 9

Low switching frequency operation of multilevel converters for high-power applications

Zahoor Ahmad Ganie[1,2], *Abdul Hamid Bhat*[1] *and Salman Ahmad*[2]

This chapter explains the low switching frequency operation of multilevel converters for high-power applications with a focus on selective harmonic minimization for controlling the harmonic magnitude from the output waveform. The intelligent solving techniques have been employed to obtain the optimal switching angles that will provide desired fundamental component and control on selected harmonics component magnitude. The insights provided in this study will benefit the researchers and engineers working on high-power application of multilevel converter systems.

9.1 Introduction

The energy generated from renewable energy sources produces DC electricity, whereas the transmission of the electric power and the loads use AC power. Therefore, a conversion device is required for converting DC power into AC power [1]. A power electronic device called inverter [2] converts DC power/voltage into AC power/voltage [3]. Such equipment or loads which need AC power require inverters for their operation [4]. Inverters are energy conversion devices that are cost-effective, simple, and efficient in operation. Inverters are often used in renewable energy systems [5] and power electronics-based sectors [6], academics and industry researchers are constantly focusing on inverter research and development. Inverters have various advantages, however, the output voltage generated from an inverter contains a fair amount harmonic content [7], which can affect the various components of the system. Lower order harmonics are more threatening to the systems than the harmonics of higher order due to the fact that these harmonics' frequency is around the fundamental frequency and have a larger magnitude [8].

[1]Department of Electrical Engineering, National Institute of Technology Srinagar, India
[2]Department of Electrical Engineering, Islamic University of Science and Technology, India

Harmonics generated cause higher switching loss in power electronic switches [9], decreasing the system efficiency and performance. The harmonics in induction motors lead to torque pulsations and speed ripples [10], hence affecting the system's lifetime and reliability. When the electrical grid receives power, the lower order harmonics present in the output voltage of the inverter cause unwanted issues in the distribution system. Multilevel inverters (MLIs) are preferred for high-power, medium-voltage applications [11] due to various advantages, like high operating voltage capacity with reduced voltage power electronic devices, high quality output voltage, least distortion of currents, smaller size filters, low dv/dt stress, smaller common-mode voltages, lesser EMI, low torque ripple. Modulation based control strategies have been devised by the researchers [12] to minimize the harmonics, hence improving the effectiveness and efficacy of the inverter. Generally, high switching frequency modulation techniques like sinusoidal pulse width modulation (SPWM), phase-shifted PWM, level-shifted PWM and space vector PWM have been used for generating high quality output along with the control of the harmonic components. In SPWM, a high frequency carrier is compared with the desired fundamental component sine wave. The pulse width modulation pulses are generated based on the logic if the modulating signal is above carrier or below carrier signal. Depending on the number of phases, the modulating signals are chosen e.g. for N phase inverter, N modulating signals equally shifted by an angle of $\frac{2\pi}{N}$ are required to generate PWM pulses. The high frequency of the carrier ensures that the harmonics spectrum is shifted at that frequency and at its multiples. In order to increase the dc bus utilization, a third harmonic component (for three phase system) of suitable peak magnitude is added throughout the modulation signal and then the resultant modulating signal is compared with the high frequency carrier signal. The third harmonics injected get canceled in the output voltage and thus will not affect the performance. The optimum value of k for maximum utilization of dc bus is $\frac{1}{6}$, although its magnitude lies in $k \in [0.15, 0.2]$ and to achieve different performance parameters its value may be taken accordingly. In phase-shifted PWM for L level configuration, the number of carriers required is $N_c = L - 1$. The shifting in carrier phases are, $\frac{360}{N_c}$. For a 7-level configuration, it is found that the three carriers are independent whereas the next three carriers are mirror image of the first three carriers. It eases the logic design when implementing in the controller. For a multiphase multilevel system, the carrier remains the same whereas the modulating signals are phase shifted by an angle of $\frac{2\pi}{N}$, where N is the number of phases. In level-shifted PWM the number of carriers is $N_c = L - 1$, which are in phase but shifted in magnitude. The device switching frequency is different for all devices in level shifted unlike the phase system case so swapping is required to manage the thermal issues. Also, unlike the phase shifted case the conduction time is different in this case [13]. The space vector pulse width modulation (SVPWM) is one of the widely used high switching frequency-based digital control techniques for converters. In a three-phase converter, if the summation of three-phase voltages is zero (balanced system), the three phase voltages can be represented by equivalent two-phase system in $dq0$ plane. The three-phase voltages are decomposed into vectors based on the switching states. If we consider '1' state when top switch of a leg is ON and '0' state when the top switch of a leg is OFF, then a total of $2^3 = 8$, switching states will be obtained. Only six switching states will be active states while the remaining two states will be zero states (000) and (111). The hexagon thus obtained can be divided into six sectors, and based on the reference vectors length and position the nearest vectors are selected with time averaging. The reference vector rotates in space with the frequency of the fundamental component. There are six switching in each sector (minimum possible). Similarly, for three level and higher level inverters nearest vector can be identified in which the reference vector falls and the dwell time can be computed. The three level have four smaller triangles in each sector (total six sectors), and all the active and zero vectors are applied based on the location and magnitude of the reference vector. SVPWM technique has been found to be a more efficient switching method in terms of power quality and better utilization of the DC bus voltage.

Selective harmonic elimination pulse width modulation (SHE-PWM) [14] approach is utilized for eradicating the lower order harmonics from the output voltage of the inverter due to its improved control and it outperforms the other developed strategies [15]. Fourier series expansion is used in SHE-PWM approach for the decomposition of the output voltage waveform [16]. Waveform features including the amplitude of the voltage level and symmetry of the waveform play a crucial part in the analysis and also affect the shape and solution space complexity [17]. Highly complex and non-linear transcendental equations are utilized in the SHE-PWM technique to generate the optimum switching patterns to remove the undesired harmonics from the inverter output voltage. Due to cosine and sine math functions, it is very difficult to achieve convergence towards optimal solution. Also, with the surge in the number of switching angles and the level of the inverter, the problem gets more serious. The main challenge is to find a meticulous solution for the SHE-PWM waveform, and the choice of an appropriate solution strategy is significantly influenced by the waveform's formulation [18]. For the calculation of switching angles for different SHE-PWM waveforms, a number of solving techniques have been proposed, like iterative techniques [19,20] algebraic methods, and intelligent optimization algorithms. The numerical technique is a highly efficient iteration technique, but its performance is mostly reliant on the initial guess. Selecting an initial guess properly in numerical techniques converges the solution very fast and finds all the possible solutions. Newton–Raphson technique is mostly used for solving the SHE equations [21,22]. SHE equations have also been solved using Walsh functions and Homotopy Algorithms numerical techniques [23–25]. The limitation of the numerical approach is that it is highly dependent on a good initial guess, hence, incorrectly chosen guesses lead to long repetitive cycles that can diverge in extreme circumstances. In algebraic methods, the initial guesses are not required and can determine all the potential solutions. In these methods, non-linear transcendental equations are translated into polynomial equations. These polynomial equations are transformed into a triangular form using the theory of resultants [26] and can be solved to calculate the optimum switching instants. These methods have very high computational time and are effective only when the switching angles are less than or equal to five. Groebner-based theory [27] and Wu-method [28] based on algebraic methods have also been utilized to determine the optimum switching instants. Calculation of real-time solutions with algebraic methods is computationally very difficult, and the difficulty increases further with the surge of levels of the inverter. Intelligent optimization algorithms are preferred to elucidate non-linear transcendental equations. These algorithms are advantageous because they do not rely entirely on initial assumptions and are not computationally complex. Genetic algorithm (GA) [29,30], Bee algorithm (BA) [31], particle swarm optimization (PSO) [32,33], ant colony optimization (ACO) [34], and differential evolution (DE) [35–37] are most commonly used optimization techniques. An objective function based on the non-linear transcendental equations of fundamental and harmonics of the lower order is used in these optimization techniques. The objective function is minimized to obtain optimal switching patterns to eliminate undesirable harmonics. The execution of the optimization technique is highly influenced by the design of an objective function.

Multilevel converters generate high voltages and high-power levels without transformers due to their unique structure. The main aim of the multilevel inverter is to generate a proper output voltage from different DC voltage levels. In a multilevel inverter as the number of levels increase the output waveform generated produces more steps, hence producing a waveform of staircase that matches the desired waveform. The harmonic distortion reduces as more and more steps are added to the waveform hence, becomes almost zero with the surge in the number of levels. Multilevel inverter topologies classification is given in Figure 9.1. The different topologies of multilevel inverters used are neutral point clamped multilevel inverter (NPC-MLI), flying capacitor multilevel inverter (FC-MLI), cascaded H-bridge multilevel inverter (CHB-MLI), active neutral point clamped multilevel inverter (ANPC-MLI), modular multilevel inverter (MMI), and packed U-cell multilevel inverter (PUC-MLI). In recent years, several new topologies are proposed. The promising topologies are T-type inverter, cascaded half-bridge-based multilevel DC link (MLDCL) inverter, switched series/parallel sources (SSPS)-based

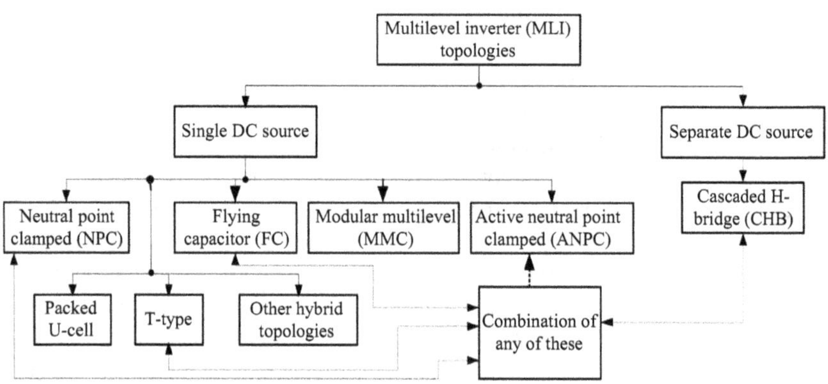

Figure 9.1 Multilevel inverter topologies

MLI, cascaded bipolar switched cells (CBSC)-based MLI, reversing voltage (RV) topology, multilevel module (MLM)-based MLI, two-switch-enabled level-generation (2SELG)-based MLI, and packed E-cell topologies. All these topologies have been designed with minimal switching device counts, a maximum number of levels, and an application-oriented approach in mind. The remaining chapter is organized as follows: Section 9.2 explains the selective harmonic minimization problem formulation, Section 9.3 explains various solving techniques, results and discussion are given in Section 9.4, in Section 9.5 comparison of different optimization techniques is given and finally in Section 9.6 conclusion and future work of the chapter is given.

9.2 Selective harmonic minimization problem formulation

Fourier series analysis of the output voltage waveform is used in SHE-PWM [13, 38] as given by (9.1), to determine the switching patterns for controlling the harmonics:

$$U(t) = \frac{1}{2}a_0 + \sum_{k=1}^{\infty} a_k \cos(k\omega t) + \sum_{k=1}^{\infty} b_k \sin(k\omega t) \tag{9.1}$$

a_0, a_k, and b_k are the Fourier series coefficients and ω is the fundamental frequency component (rad/s). The value of Fourier coefficients in (9.1) are calculated from (9.2) and (9.3):

$$a_k = \frac{2}{T}\int_{-\frac{T}{2}}^{\frac{T}{2}} U(t)\cos(k\omega t) \tag{9.2}$$

$$b_k = \frac{2}{T}\int_{-\frac{T}{2}}^{\frac{T}{2}} U(t)\sin(k\omega t) \tag{9.3}$$

The time period of the waveform is given by T. The SHE-PWM problem formulation for any waveform requires quarter wave (QW) symmetry [39] due to which the DC component, sine coefficients of odd harmonics and even harmonics become zero, hence the problem formulation and solving strategy is substantially simplified. Therefore output voltage is given by (9.4)

$$U(t) = \sum_{k=1}^{\infty} b_k \times \sin(kwt) \tag{9.4}$$

where k is the harmonic order, b_k is the magnitude of kth harmonic. The value of b_k must be equated to zero for the particular harmonic to be zero. Hence, by substituting the appropriate b_k to zero, any harmonic of specific order can be eradicated from the waveform. For N switching angles per quarter cycle, (N–1) harmonics can be eradicated from the waveform [40]. The cascaded H bridge multilevel inverter is shown in Figure 9.2 and the corresponding stepped multilevel

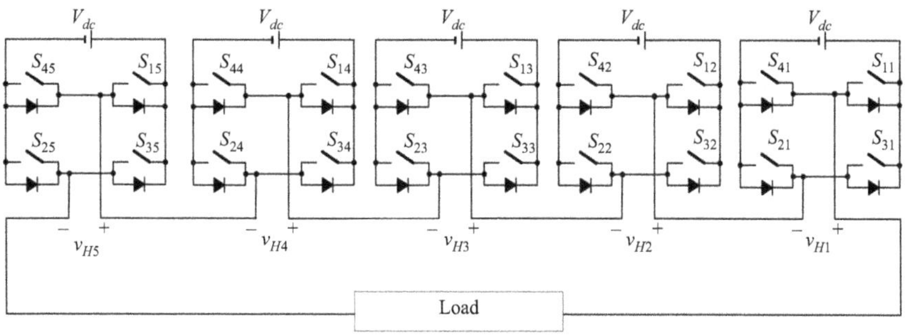

Figure 9.2 Cascaded H bridge multilevel inverter topology

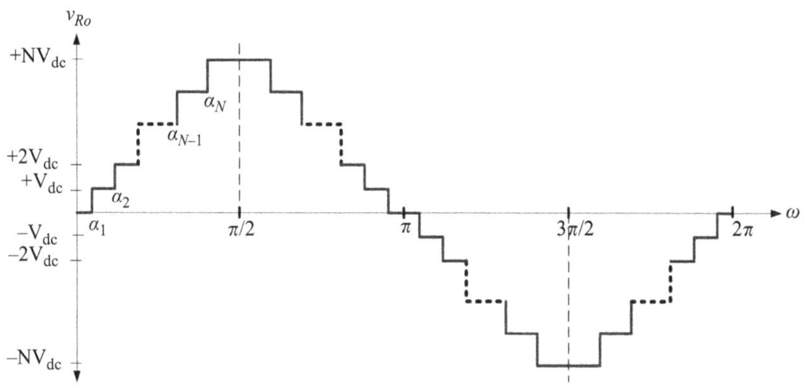

Figure 9.3 Generalized multilevel stepped waveform

waveform is shown in Figure 9.3. The magnitude of fundamental component and harmonics is calculated using (9.5):

$$b_k = \frac{4V_s}{k\pi}\left[\sum_{j=1}^{N}(-1)^l \times \cos\left(k\alpha_j\right)\right] \tag{9.5}$$

where V_s represents the DC source voltage and N represents the total number of switching angles. The SHE-PWM problem formulation for multilevel waveform can be stated by the number of shifts that occur in one time frame, and the dispersal of changes among the waveform's multiple levels. The distribution of switching angles between levels makes multilevel SHE-PWM formulations more complex and has an impact on solution convergence and consistency. Waveform transition is defined by the parameter l, whose values are given by (9.6)

$$l = \begin{cases} 0 ... for .. rising .. edge \\ 1 ... for .. falling .. edge \end{cases} \tag{9.6}$$

The switching angles in the quarter period are restricted between $0 \leq \alpha_1 \leq \alpha_2 \leq .... \leq \alpha_N \leq \frac{\pi}{2}$ to produce a symmetrical and accurate waveform. The fundamental frequency component V_1 is determined by the modulation index (m_a)

$$V_1 = \frac{4m_a V_s}{\pi} \tag{9.7}$$

The value of m_a is between $0 \leq M \leq 1$.

9.3 Solving techniques

SHE-PWM is utilized to determine the optimum switching patterns from the non-linear transcendental equations to remove the specific harmonics from the output voltage waveform. Various SHE-PWM approaches have been unfolded in the literature to determine the optimal value of switching angles. The different solving techniques have been developed which are categorized as

1. Numerical techniques (NMs)
2. Algebraic methods (AMs)
3. Intelligent algorithms (IAs)

9.3.1 Numerical techniques

The NM are iterative approaches to determine the optimum solutions in the minimum possible time. Nonetheless, these approaches are reliant on the starting values and become entangled at the local minimal solution [41]. The predictive [42] and intelligent approaches [43] are used to determine the initial values for Newton–Raphson (NR). The Walsh function [23], homotopy algorithm [24], sequential quadratic programming [44], and gradient optimization [45] are some of the enhanced NMs reported in the literature. Nonetheless, these approaches are also highly reliant upon the initial guesses.

9.3.2 Algebraic methods

In these methods, the non-linear transcendental equations are translated into polynomial equations to determine the optimal switching angles. The primacy of these methods is that all the solutions are calculated without the need for any initial guesses [27]. The transformation of equations is done using Chebyshev polynomials [46] by substituting $\cos(k\alpha_o) = y_o$ and $\cos(k\alpha_e) = y_e$ where the indices o and e represent odd and even numbers, thus producing a set of algebraic polynomials as shown in (9.8)

$$\begin{cases} y_1 + y_2 + \dots + y_n = r_1 \\ y_1^3 + y_2^3 + \dots + y_n^3 = r_3 \\ \vdots \\ y_1^{2n-1} + y_2^{2n-1} + \dots + y_n^{2n-1} = r_{2n-1} \end{cases} \tag{9.8}$$

where y_1, y_2,..., y_n contain the solutions of the firing angles α_1, α_2 and α_n respectively. r_1, r_2,..., r_{2n-1} are determined employing recursive algorithm [46] as shown in (9.9)

$$H_{2n-1}(y)|_{y^{2n-1}=r_{2n-1}} = \frac{1}{2} + \frac{\pi \times b_{2n-1}}{8V_{dc}} \tag{9.9}$$

where H_{2n-1} is the Chebyshev polynomial and its value is calculated using (9.10)

$$H_{n+1}(y) = 2yH_n(y) - H_{n-1}(y) \tag{9.10}$$

where $H_o(y) = 1$ and $H_1(y) = y$. For a three-level inverter, the value of r is calculated using (9.11)

$$r_{2n-1} = \frac{\pi \times b_1}{4^{n-1} \times 4V_{dc}} \binom{2n-1}{n-1} \tag{9.11}$$

Since the algebraic equations are the symmetric sum of powers, these equations are then converted into a single polynomial using Newton's identities as shown in (9.12)

$$P(x) = p_o y^n + p_1 y^{n-1} + \ldots + p_n = 0 \tag{9.12}$$

where $p_0 = 1$, and the value of p_1 to p_n are calculated using Newton's identity.

Resultant theory [47], Wu technique [28], symmetric polynomial theory, Groebner-based theory [27], and Power sum [48] have also been utilized for harmonic elimination in inverters. However, these solutions are very complex computationally and are not suitable for real-time applications of inverters.

9.3.3 *Intelligent algorithms*

Artificial intelligence approaches motivated by nature are called Bio-Intelligent Algorithms (BIAs) [49,50]. These algorithms are population-based iterative procedures where starting values have very little impact on the optimization process. These algorithms are easy to learn and programmed using any computing software on PCs. An objective function is used in these algorithms which includes low-order harmonics and fundamental transcendental equations. The objective function is minimized by these algorithms to calculate the optimal switching patterns. The effectiveness of these algorithms is highly influenced by the design of an objective function. The objective function given by (9.13) is employed in this investigation:

$$OF = \left(100 \times \frac{V_1^* - V_1}{V_1^*}\right)^4 + \sum_{k=5,7,..}^{N-1} \frac{1}{k}\left(50 \times \frac{V_k}{V_1}\right)^2 \tag{9.13}$$

where k denotes the harmonic order. The objective function's initial component is fine-tuned to a power of 4, reducing the error between V_1 and V_1^* to less than 1%. For violations with a margin of error of less than 1%, this power has negligible effect. The undesired lower order harmonics are confined to less than 2% of the fundamental, therefore is fined tuned with a power of two.

The most popular intelligent algorithms are described in detail in the next section.

9.3.3.1 Genetic algorithm (GA)

John Holland's University of Michigan students and colleagues [51] developed the GA which is an adaptive heuristic search strategy. Charles Darwin's "survival of the fittest" hypothesis is the basis of this algorithm. The GA is an optimization algorithm used for solving various optimization problems without constraints. The algorithm starts with the collection of chromosomes known as "population" which holds problem-solving answers. An objective function is utilized to ascertain every chromosome's fitness in this algorithm. The fit parents are chosen depending on their fitness and crossover operation is employed to create children for the upcoming generation. The fittest chromosomes are chosen as parents in every iteration to produce children which inherit the characteristics of their parents. These characteristics enhance the chances of attaining better results in succeeding iterations. After every iteration, the solution with the optimum fitness is noted. This procedure is replicated till the stopping norms are met. The possible solution sets are encoded in a binary string of '0' and '1' of length $'l'$ as given by (9.14):

$$\zeta_1 = [11101...01011], ...\zeta_2 = [10100...11010], ...\zeta_m = [01001...00111] \tag{9.14}$$

Here the set of $\{\zeta_1, \zeta_2...\zeta_m\}$ are called chromosomes and ζ_i as genes. The population set is calculated using (9.15):

$$Pop = \left\{ \begin{array}{c} \zeta_{1,1}, \zeta_{2,1}, \cdots \zeta_{m,1} \\ \zeta_{1,2}, \zeta_{2,2}, \cdots \zeta_{m,2} \\ \cdot \\ \cdot \\ \cdot \\ \zeta_{1,X}, \zeta_{2,X}, \cdots \zeta_{m,X} \end{array} \right\} \tag{9.15}$$

The fitness function of every chromosome is evaluated and these fitness function values are appended to calculate the total fitness value. For selecting chromosomes, the fitness values are divided to obtain weight/probability p_i. Offsprings are generated using the crossover procedure. The GA flowchart for calculating the switching angles, total harmonics distortion, and harmonic profile is shown in Figure 9.4.

9.3.3.2 Particle swarm optimization (PSO)

PSO algorithm was first developed by Eberhart and Kennedy [52]. This algorithm is population-based and motivated by the schooling of fish and the swarming of birds while they look for food. At the start, the random population of particles is generated. For N variables, a swarm particles of N_p are defined and each particle has any random position in n-dimensional space. The trajectory

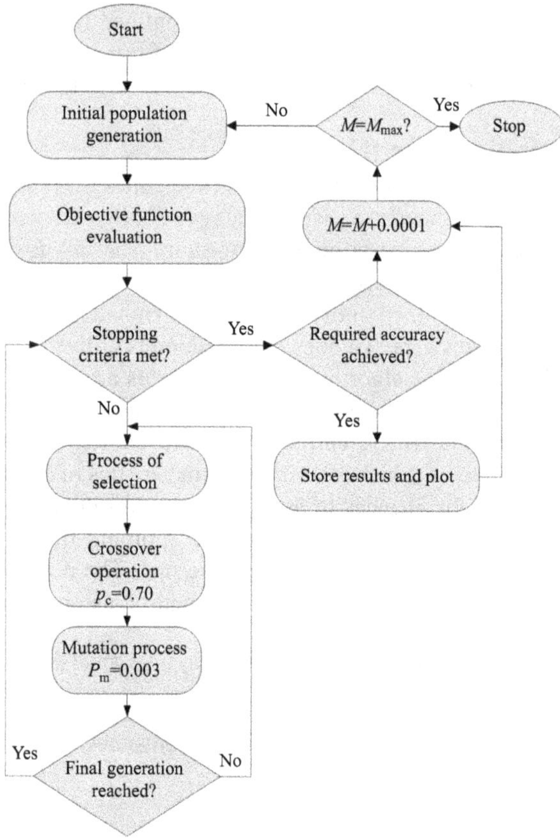

Figure 9.4 Genetic algorithm flow chart

(position and velocity) x_k and V_k of every particle are defined. All the particles upgrade their trajectories depending upon their best position and the trajectories of other particles. All of this is achieved by optimizing an objective function. The particles best P_{best} and global best G_{best} are upgraded in every iteration. At $(i+1)$th iteration the swarm are updated using (9.16) and (9.17) respectively [33]:

$$V_k(i+1) = V_k(i) + d_1 s_1 \left[P_{best,k} - x_k(i)\right] + d_2 s_2 \left[G_{best,k} - x_k(i)\right] \tag{9.16}$$

$$x_k(i+1) = V_k(i+1) + x_k(i), \; k = 1, 2, ..., N_p. \tag{9.17}$$

Here, d_1 and d_2 are the cognitive learning rate and s_1 and s_2 are the social learning rates having uniform distribution within $[0, 1]$. The PSO flowchart for calculating the switching angles, total harmonics distortion, and harmonic profile is shown in Figure 9.5.

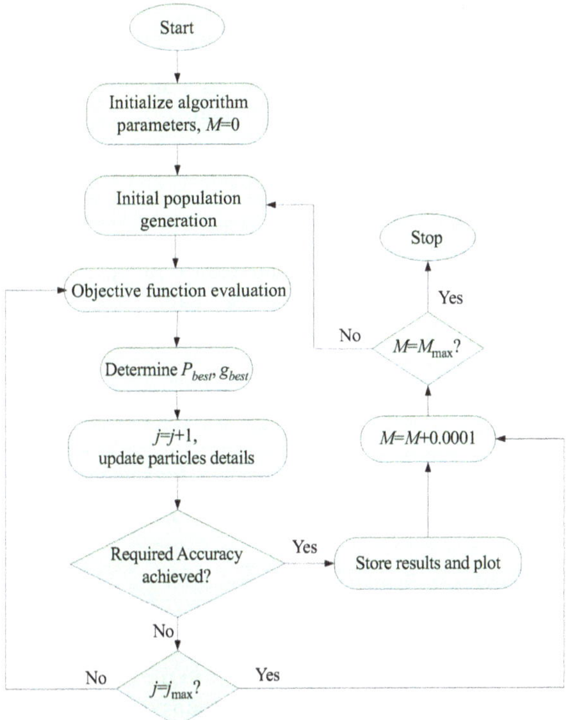

Figure 9.5 PSO flow chart

9.3.3.3 Bee algorithm (BA)

Karaboga and Yang developed BA [53]. The foraging behavior of honeybees inspired this algorithm. This algorithm is split into three stages: (i) the forager bee stage, (ii) the observer bee stage, and (iii) the scout bee stage. Using the objective function given by (9.18), the fitness of every food source is computed. Employed bees search for fresh food sources s_{mn} in the vicinity of y_{mn} in each iteration and if better sources are found the food sources are updated:

$$s_{mn} = y_{mn} + \phi(y_{mn} - y_{kn}) \tag{9.18}$$

ϕ is a random number whose value lies between [−1, 1]. At the end of the employed bee phase, observer bees are provided with the best information about food sources. Bees in this phase are directed to the areas having large amount of nectar found by employed bees. The onlooker bees explore for fresh food sources s_m around the area y_m. If the sources are found to have a greater supply of nectar than the present food sources, the former will be replaced by the latter. In every iteration collaboration between bees from both phases results in better solutions. If certain sources of food does not improve after a specific number of iterations, these sources of food will be randomly changed by scout bees. This process is carried out till the termination norms are met.

9.3.3.4 Differential evolution (DE)

DE was developed by Storn and Price [54]. This method uses population-based optimization to determine the best solution possible for complicated problems iteratively. This algorithm uses an objective function to assess each agent's fitness. For every member in the population, a mutant vector M is produced as shown in (9.19) by randomly choosing three individuals J, K, and L from the population:

$$M = J + m_f(K - L) \tag{9.19}$$

where m_f is the mutation factor chosen from the range [0, 2]. The crossover procedure of target vector y_j and mutant vector M produces a child vector C_v. The trial vector T_v is produced using the probability of crossover rate R_{co} given by (9.20)

$$T_v = \begin{cases} C_v..if.rand. \leq R_{co} \\ y_j..if.rand. \geq R_{co} \end{cases} \tag{9.20}$$

R_{co} has the range [0, 1]. Whenever the fitness of trial vector T_v is lesser than the target vector y_j then the target vector y_j is replaced by the trial vector T_v as shown in (9.21)

$$y_j = \begin{cases} T_v..if..f(T_v) \leq f(y_j) \\ y_j..if..f(T_v) \geq f(y_j) \end{cases} \tag{9.21}$$

This strategy is replicated for every member in the population. The program is terminated as soon as the number of iterations attains the maximum or the value of the objective function reaches the required level. The DE algorithm flowchart for calculating the switching angles, total harmonics distortion, and harmonic profile is shown in Figure 9.6.

9.3.3.5 Imperialist competitive algorithm (ICA)

Atashpaz-Gargari and Lucas were the first to introduce ICA [55]. Imperialistic rivalry motivated this algorithm. In this algorithm, the population is referred to as "countries," that are split up between various empires. Every empire is made up of imperialist and a number of colonies. The rivalry among empires results in trouncing the colonies of fragile empires by powerful empires. An empire crumbles due to the loss of colonies. Finally, a single empire with one imperialist rules all the colonies. Every country's fitness is determined by utilizing an objective function. The countries are then arranged in increasing order after determining their fitness values and the first m countries are chosen as imperialist imp_m. For the division of colonies normalized cost C_m and power P_m of every imperialist is determined using (9.22) and (9.23), respectively:

$$C_m = c_m - \max\left[c_j\right] \tag{9.22}$$

$$P_m = \left[\frac{C_m}{\sum_j^m C_j} \right] \tag{9.23}$$

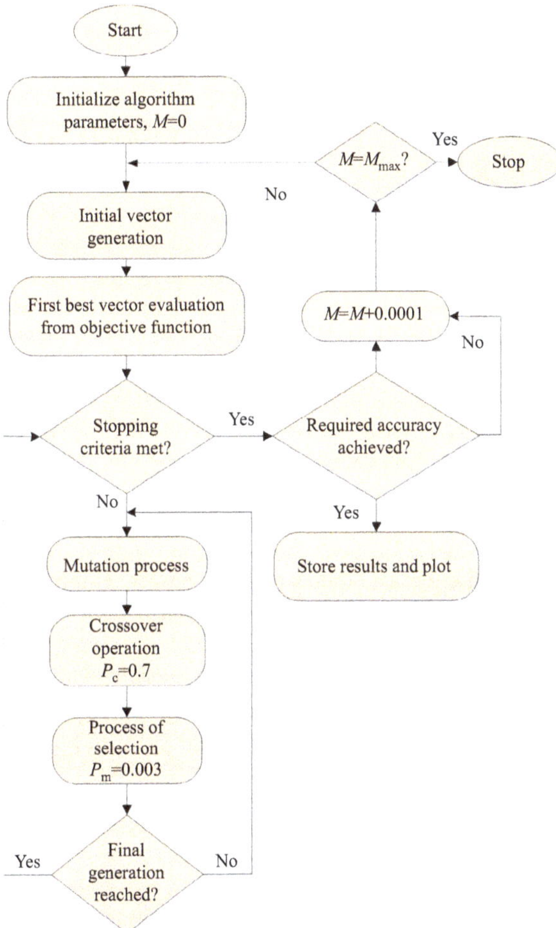

Figure 9.6 DE algorithm flow chart

Colonies are divided using (9.24) depending upon the normalized power of every imperialist

$$M \times C_m = round[P_m \times M_{col}] \tag{9.24}$$

Colonies in the empire move closer to their appropriate imperialist in every iteration using (9.25)

$$z_{new} = z_{old} + \alpha \times rand \times [z_{imp} - z_{old}] \tag{9.25}$$

where α is a constant number ≥ 1, *rand* represents a random number whose value lies between [0, 1] with uniform distribution. An empire's overall power is determined using (9.26):

$$T \times C_m = Cost(imp_m) + \beta mean\,[cost(colonies_m)] \tag{9.26}$$

where β has a positive value but less than one, and $colonies_m$ represents the colonies of empire m. To guarantee unbiased competition normalized total cost of every empire is determined using (9.27)

$$M \times T \times C_m = T \times C_m - \max(T \times C_j) \tag{9.27}$$

$T \times C_m$ gives the total cost of mth empire. The probability of possession of every empire is calculated using (9.28):

$$P_{P_m} = \left[\frac{M \times T \times C_m}{\sum_j^m M \times T \times C_j} \right] \tag{9.28}$$

For handing over the weak colonies to the strongest empire the probability of possession is arranged in the form of a vector, as shown in (9.29):

$$P = [P_{P_1}, P_{P_2}, P_{P_3}, \ldots, P_{P_m}] \tag{9.29}$$

Then, a vector B is generated as shown in (9.30) with the same dimensions as P, containing random numbers between [0, 1] with uniform distribution:

$$B = [b_1, b_2, b_3, \ldots, b_m] \tag{9.30}$$

Vector E is created using (9.31)

$$E = P - B \tag{9.31}$$

A fragile colony is taken over by the empire having the highest relevant index in E. This process is repeated in every iteration resulting in the collapse of a fragile empire and a strong empire taking over its colonies. Empires without colonies are eliminated from the contest. At last, a single empire survives conquering all the colonies. A powerful empire's imperialist and its colonies have the same objective function value, implying that they are identical. The program will be stopped when the most powerful empire has captured all of the colonies, or when the number of iterations has reached the maximum.

9.3.3.6 Firefly algorithm (FA)

FA was developed by Yang [56]. The sparkling behavior of fireflies motivated this algorithm. This algorithm is adaptable, easy and straightforward to put into action. The ferocity of the light of all fireflies in this algorithm is determined using the objective function. Fireflies with lesser intensity move towards the fireflies with brighter intensity in the swarm depending upon the value of light intensity. Firefly with less intensity y_j moves towards the firefly with more intensity y_k using (9.32)

$$y_j^{t+1} = y_j^t + \alpha_0 e^{-\beta r_{jk} m}\left(y_k^t - y_j^t\right) + \gamma\left(rand - \frac{1}{2}\right), m \geq 1 \tag{9.32}$$

Firefly's current position is given by y_j^t, attractiveness at distance $r = 0$ is given by α_0, search space exploration and exploitation are controlled by randomization parameter γ, and the coefficient of light absorption is β which controls the velocity and behaviour of this algorithm.

9.3.3.7 Ant colony optimization (ACO)

ACO is a meta-heuristic technique that utilizes the cooperation among ants to find the shortest path in order reach to a food location to minimize the objective function [57]. This algorithm is mostly used for discrete problems and produces a rapid solution. The steps involved in this algorithm are initialization, and random solution guided from pheromone trails. The optimal solution is one where all the ants converge. The probability of an ant 'n' moving from node j to k in generation i is given by (9.33):

$$P_{j,k}^n(i) = \frac{\Phi_{j,k}(n)\, d_{j,k}^{-\mu}}{\sum_{y \in \psi_j^n} \Phi_{j,y}\, d_{j,y}^{-\mu}}, \quad m \in \psi_j^n \tag{9.33}$$

$\Phi_{j,k}$ is the pheromone intensity at $j \rightarrow k$, $d_{j,k}$ is the route from node j to k, and ψ_j^n is the remaining nodes for the visit. The record of ant visits is maintained in a tabu list and it keeps on updating. The expression in (9.34) is used to update the record:

$$\Phi_{j,k}(i+1) = (1-\varsigma)\Phi_{j,k}(n) + \sum_{n=1}^{N} \Phi_{j,k}^n(i) \tag{9.34}$$

$\phi_{j,k}^n(i)$ is the pheromone intensity and ς is pheromone decay parameter. The value of $j \rightarrow k$ is $\frac{Q}{L_i}$ if ant n walks in it, else it is 0. The pheromone on a particular path is renewed using (9.35):

$$\gamma_{l,m}(n+1) \leftarrow \max\left[\gamma_{\min}, \gamma_{l,m}(n+1)\right] \forall(l,m) \tag{9.35}$$

9.3.3.8 Cuckoo search algorithm (CSA)

CSA was developed by Deb and Yang [58]. This algorithm is motivated by the natural behavior of cuckoo species and fruit flies Levy flight behavior. The main steps of this algorithm are:

1. Each cuckoo only ever produces one egg at a specific moment in any nest.
2. Superior eggs from the best nest are passed down to the future generation.
3. The host nests at disposal are fixed, and a host bird recognizing the cuckoo eggs probability is $p_a\varepsilon(0,1)$. If a host bird recognizes foreign eggs, either throw away the eggs or disown the nest and build a new one elsewhere.

Solutions are represented by the eggs in the host nest, with the cuckoo egg representing a new solution. Nonetheless, only high-quality solutions will survive for the next generation. Cuckoo egg is generated using (9.36). The cuckoo egg's fitness is compared with a randomly chosen egg. If the cuckoo egg's fitness outperforms the selected egg, the host nest solution replaces the cuckoo solution. An inferior nest is rejected and new solutions are produced after every iteration.

$$x_i^{t+1} = x_i^t + \alpha + Levy(\zeta) \tag{9.36}$$

α represents the step size whose value depends on the scales of the problem under consideration. Levy flight, which is accountable for arbitrary walks, is denoted by the term Levy. For the scrutiny of the search space, Levy flight is very important, and it is calculated using (9.37):

$$Levy_n = t^{-\zeta}; ..1 \leq \zeta \leq 3 \tag{9.37}$$

9.3.3.9 Bat optimization algorithm (BOA)

BOA was proposed by Yang [59]. This algorithm is motivated by the ultrasonic behavior of bats. The ultrasonic behavior of bats assists them in locating a target. Bats emit loud sound pulses and depending upon the echo that is returned back make decisions. Bats can detect the class of object by its distance and background using echo information. Bats arbitrary fly at position y_j with fixed frequency $f_{\min}$, speed v_j, loudness L_0, and varying wavelengths λ to search their food or target. Bats modify the frequency or wavelength of their pulses automatically and control the pulse rate of emission $r\varepsilon[0, 1]$ depending on the information received from the target. This algorithm employs the same idea and terminology as used in PSO for speed and position. The speed and position of each bat are determined using (9.38) and (9.39), respectively

$$v_j^{t+1} = v_j^t + \left(v_j^t - y_j^*\right) f_j \tag{9.38}$$

$$y_j^{t+1} = y_j^t + v_j^{t+1} \tag{9.39}$$

y_j^* is the current global best position and f_j is the fixed frequency which is calculated using (9.40):

$$f_j = f_{\min} + (f_{\max} - f_{\min})\gamma \tag{9.40}$$

γ represents a random number with uniform distribution whose value lies in the range of [0, 1], $f_{\min}$ represents the minimum frequency range and $f_{\max}$ represents the maximum frequency range. The term f_j controls the pace and range of bat movement.

9.3.3.10 Invasive weed optimization (IWO)

The IWO algorithm was developed by Mehrabian and Lucas in 2006 [60]. This algorithm is a meta-heuristic optimization technique based on the ecological behavior of the weeds. The IWO algorithm is relatively simple and effective in finding the best possible solutions. This algorithm is started by the initialization of the population of invasive weed plants and randomly spreading them over the D-dimensional space. The search space dimensions are determined by the number of variables selected. The initial random population is calculated using (9.41)

$$\beta_j = \left(\beta_j^{\max} - \beta_j^{m} in\right) \times rand(0,1) + \beta_j^{\min} \tag{9.41}$$

where, $\beta_j^{\max}$ is $\pi/2$ and $\beta_j^{\min}$ is 0 for quarter wave symmetrical waveform. The rank of the weeds in the weed colony is determined by the fitness value of the weeds. Every weed is able to produce new offspring according to its fitness. The seeds produced by the *m*th weed are calculated using (9.42)

$$N_m = \frac{M - M_{\min,k}}{M_{\max,k} - M_{\min,k}} \times (D_{\max} - D_{\min}) + D_{\min} \tag{9.42}$$

N_m represents the number of off-springs from the *m*th parent, M is the fitness of *m*th parent, $M_{\min,k}$ is the minimum value of fitness and $M_{\max,k}$ is the maximum value of fitness of the parents. $D_{\min}$ is the minimum number of seeds generated by *m*th parent, $D_{\max}$ is the maximum number of seeds generated by *m*th parent. The seeds generated are randomly distributed over the D-dimensional space with zero mean and varying variance. The standard deviation (SD) of the random function will change from an initial value $\alpha_{initial}$ to a final value α_{final} in every step. The current SD can be calculated using (9.43)

$$\alpha_{SD} = \left(\frac{I_{\max} - I}{I_{\max}}\right)^n \times (\alpha_{initial} - \alpha_{final}) + \alpha_{final} \tag{9.43}$$

where I represents the number of iterations, $I_{\max}$ represents the maximum number of iterations, and n represents non-linear modulation index. After certain iterations, the number of weeds in a weed colony will attain its peak, then every weed is permitted to generate seeds according to (9.43). These seeds will roam over the search area initially and after taking their respective positions in the search area they are grouped alongside their parents. Weeds with poor fitness values are eliminated. The fittest ones are allowed to reproduce. For competitive elimination, the control mechanism is likewise applied to their off-springs till the end of a given run. This algorithm has been used in [61] for harmonics minimization in standalone PV systems.

9.4 Results and discussion

The 11-level stepped waveform is computed using various optimization techniques discussed in the previous section for five switching angles. As a case study, the computational results for 11-level stepped waveform using a particle swarm optimization algorithm have been shown here. The solution trajectories with many solutions in some portion of the modulation index are shown in Figure 9.7(a). The harmonics spectrum at $M_1 = 0.69$, $M_2 = 0.835$ and $M_3 = 0.847$ is shown in Figure 9.7(b). It is apparent from

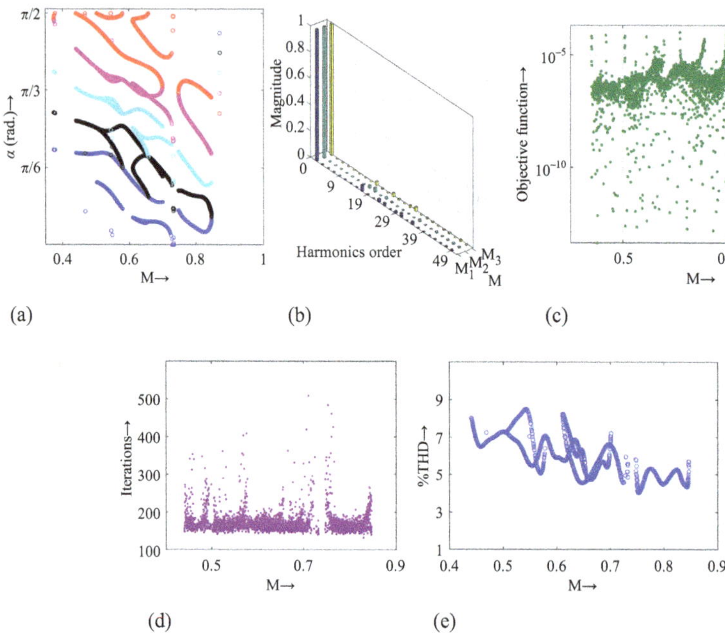

Figure 9.7 Computational results for 11-level cascaded waveform: (a) switching angles, N = 5, (b) harmonics profile, (c) objective function values, (d) number of iterations, and (e) total harmonics distortion

the harmonics spectrum that the intended harmonics considered for minimization are having negligible magnitude. The objective function value and iterations as a function of modulation index are shown in Figure 9.7(c) and (d) respectively. The value of the function is nearly 10^{-6} and the mean iteration of convergence lies between 150 and 200. The %THD as a function of modulation index is also shown in Figure 9.7(e).

Similarly for other levels, the solution angles trajectories can be found. From various algorithms tested, it was found that the evolutionary algorithm is better for addressing bipolar waveforms, however, the method takes a longer time to compute the switching angles in multilayer situations. The PSO is the most ideal scheme for addressing multilevel inverter instances, and its accuracy is the best of all the methods presented here. The accuracy is the same as that of algebraic approaches presented in the literature. This technique can also be used to analyze a large number of switching angles, which is a significant benefit. Although the DE approach has a high convergence rate, the objective function values and harmonics examined for minimization are significantly higher than those of the GA and PSO algorithms. In the continuous domain, the ACO algorithm has a low convergence rate and minimized objective function values. Many other new optimization techniques, including simulated annealing, firefly algorithm, cuckoo search, artificial bee algorithm, colonial competitive algorithm, and so on, can be used to compute switching angles, and their performances can be compared in terms of accuracy, convergence rate, and computation time.

9.5 Comparative analysis

The comparison of different optimization techniques in terms of convergence rate is shown in Figure 9.8. The choice of objective functions also influences computing time, solution correctness, and solution throughout the modulation index range. The algorithm parameters such as the number of iterations, population size, fitness function value, convergence rate, and other parameters for PSO, GA, and DE are shown in Table 9.1. The switching angles trajectories, harmonics profile, and total harmonics distortion are found to be approximately the same for all the optimization techniques. The objective function is carefully set for many switching scenarios since the change in sign influences the computational output. These intelligent approaches have several drawbacks, such as being very sensitive to input parameters, and so on. One critical issue for the real-time implementation of both numerical and intelligent methods is that, because prior knowledge regarding the understanding of the solution is not known, when these techniques fail to produce a result, it is not entirely clear whether this is due to the initial value guess, the parameters being insufficient, or there are no solutions to the SHE equations. Algebraic approaches, which do not require initial

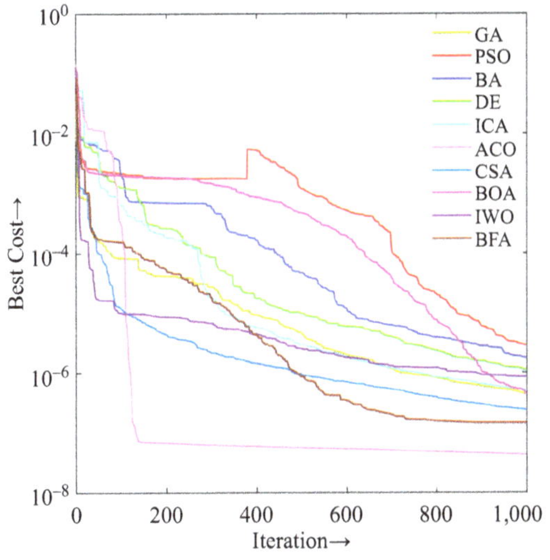

Figure 9.8 Performance comparison of optimization algorithms

Table 9.1 Algorithm parameters for PSO, GA and DE

Algorithm	Iterations	Population	Fitness	Convergence	Other parameters
PSO	300	150	1×10^{-6}	Faster	$\phi_1 = 2.05; \phi_2 = 2.05$
GA	200	100	1×10^{-5}	Fast	$p_c = 0.70; p_m = 0.003$
DE	100	1,000	0.5×10^{-5}	Moderate	$P_c = 0.70; P_m = 0.003$

values and can compute all solutions, can overcome the constraints of numerical and intelligent methods. The SHE equations are translated into polynomial equations using the variables substitution and multiple angle formulae and these polynomial equations are then transmuted into their equivalent triangular form using resultant elimination theory, which is readily solved. This approach has low efficiency and is only useful when the DC sources or switching angles are less than four.

The experimental setup for the cascaded multilevel inverter to verify the low switching frequency PWM concept is shown in Figure 9.9. A DSP controller is used to produce the gate pulses for the gate driver board of the multilevel converter. The switching tables are stored in the CPU as LUTs, and the processor selects the appropriate table based on the needed output voltage. The experimental result for 11-level waveform with RL load (R = 10 Ω, L = 3 mH) is shown in Figure 9.10.

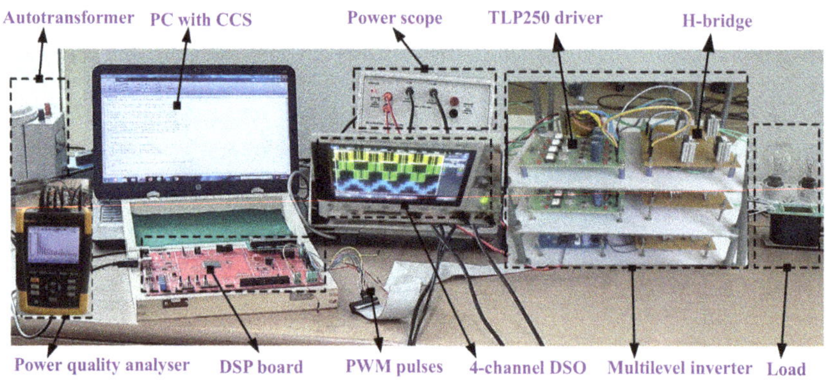

Figure 9.9 Hardware setup

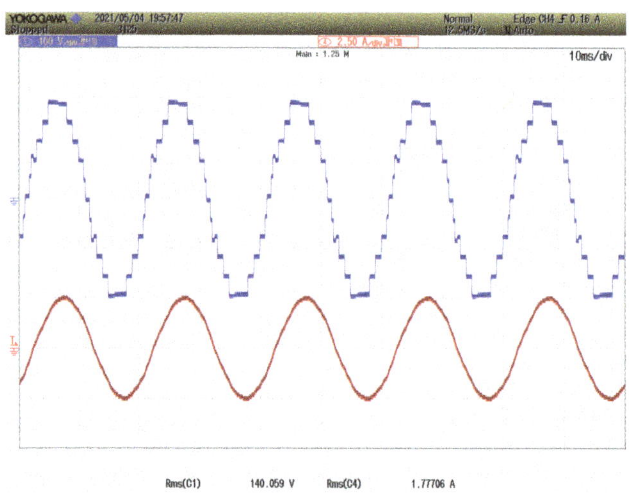

Figure 9.10 Hardware result for 11-level CHB with RL load

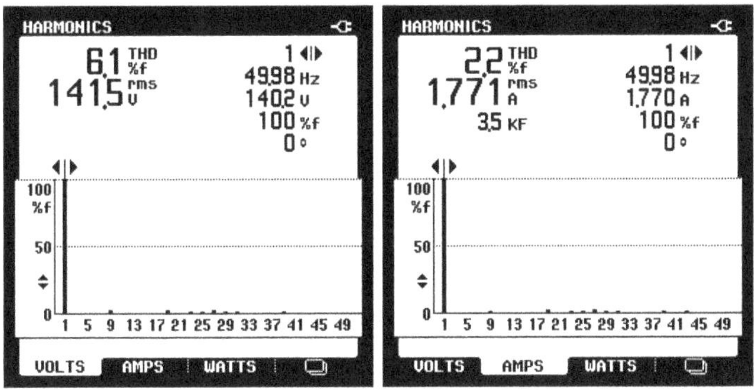

(a) Voltage harmonics profile, N=5 (b) Current harmonics profile, N=5

Figure 9.11 Harmonics profile of voltage and currents

The current waveform is more like a sinusoidal waveform because of the nature of the load. The harmonics profile for voltage and current waveforms is shown in Figure 9.11. It is clear that the intended harmonics for minimization have negligible magnitude. Moreover, the harmonics profile of the current waveform has less total harmonics distortion than the voltage waveform because of inductive load.

9.6 Conclusion and future work

In this chapter, a low switching frequency operation of multilevel converters by selectively controlling the harmonic magnitude from the output waveform has been discussed. Intelligent solving techniques have been employed to obtain optimal switching angles. Moreover, many metrics like as precision, computational complexity, faster convergence, and a wide range of control parameters are used to evaluate the performance of intelligent algorithms. Simulations using a 11-level cascaded H-bridge inverter are used to validate this strategy.

Future work: the real implementation of the computed switching for fast dynamic response of induction motor drives may be developed in order to save the controller memory and computational burden. Also, the online implementation of intelligent techniques will avoid the lookup table-based issues.

References

[1] Iqbal A, Moinoddin S, Ahmad S, *et al.* Multiphase converters. In: *Power Electronics Handbook.* New York, NY: Elsevier; 2018. p. 457–528.

[2] Mohan N, Undeland TM, and Robbins WP. *Power Electronics: Converters, Applications, and Design*. New York, NY: John Wiley & Sons; 2003.

[3] Ahmad S, Uddin R, Ghanie ZA, *et al.* Close loop control of quasi Z-source inverter in grid connected PV system. In: *2019 International Conference on*

Electrical, Electronics and Computer Engineering (UPCON). New York, NY: IEEE; 2019. p. 1–6.
[4] Ahmad S, Uddin R, Ganie ZA, *et al.* Low frequency operation and dsPIC micro-controller implementation for multilevel quasi Z source inverter in photovoltaic application. *Distributed Generation & Alternative Energy Journal*. 2022; p. 929–958.
[5] Panwar N, Kaushik S, and Kothari S. Role of renewable energy sources in environmental protection: a review. *Renewable and Sustainable Energy Reviews*. 2011;15(3):1513–1524.
[6] Rashid MH. *Power Electronics Handbook*. Oxford: Butterworth-Heinemann; 2017.
[7] Erickson RW and Maksimovic D. *Fundamentals of Power Electronics*. New York, NY: Springer Science & Business Media; 2007.
[8] Ahmad S, Ashraf I, Iqbal A, *et al.* SHE PWM for multilevel inverter using modified NR and pattern generation for wide range of solutions. In: *2018 IEEE 12th International Conference on Compatibility, Power Electronics and Power Engineering (CPE-POWERENG 2018)*. New York, NY: IEEE; 2018. p. 1–6.
[9] Kolar JW, Ertl H, and Zach FC. Influence of the modulation method on the conduction and switching losses of a PWM converter system. *IEEE Transactions on Industry Applications*. 1991;27(6):1063–1075.
[10] Casadei D, Profumo F, Serra G, *et al.* FOC and DTC: two viable schemes for induction motors torque control. *IEEE Transactions on Power Electronics*. 2002;17(5):779–787.
[11] Rodriguez J, Franquelo LG, Kouro S, *et al.* Multilevel converters: an enabling technology for high-power applications. *Proceedings of the IEEE*. 2009;97(11):1786–1817.
[12] Trzynadlowski AM. An overview of modern PWM techniques for three-phase, voltage-controlled, voltage-source inverters. In: *Proceedings of IEEE International Symposium on Industrial Electronics*, vol. 1. New York, NY: IEEE; 1996. p. 25–39.
[13] Holmes DG and Lipo TA. *Pulse Width Modulation for Power Converters: Principles and Practice*, vol. 18. New York, NY: John Wiley & Sons; 2003.
[14] Memon MA, Mekhilef S, Mubin M, *et al.* Selective harmonic elimination in inverters using bio-inspired intelligent algorithms for renewable energy conversion applications: a review. *Renewable and Sustainable Energy Reviews*. 2018;82:2235–2253.
[15] Ahmad S, Ganie ZA, Ashraf I, *et al.* Harmonics minimization in 3-level inverter waveform and its FPGA realization. In: *2018 3rd International Innovative Applications of Computational Intelligence on Power, Energy and Controls with their Impact on Humanity (CIPECH)*. New York, NY: IEEE; 2018. p. 1–5.
[16] Ahmad S, Meraj M, Iqbal A, *et al.* Selective harmonics elimination in multilevel inverter by a derivative-free iterative method under varying voltage condition. *ISA Transactions*. 2019;92:241–256.

[17] Ahmad S, Khan I, Iqbal A, *et al.* A novel pulse width amplitude modulation for elimination of multiple harmonics in asymmetrical multilevel inverter. In: *2021 IEEE Texas Power and Energy Conference (TPEC)*. New York, NY: IEEE; 2021. p. 1–6.

[18] Wells JR, Nee BM, Chapman PL, *et al.* Selective harmonic control: a general problem formulation and selected solutions. *IEEE Transactions on Power Electronics*. 2005;20(6):1337–1345.

[19] Ahmad S, Al-Hitmi M, Iqal A, *et al.* Low-order harmonics control in staircase waveform useful in high-power application by a novel technique. *International Transactions on Electrical Energy Systems*. 2019;29(3):e2769.

[20] Ahmad S, Iqbal A, Al-Ammari R, *et al.* Selected harmonics elimination in multilevel inverter using improved numerical technique. In: *2018 IEEE 12th International Conference on Compatibility, Power Electronics and Power Engineering (CPE-POWERENG 2018)*. New York, NY: IEEE; 2018. p. 1–6.

[21] Sun J, Beineke S, and Grotstollen H. Optimal PWM based on real-time solution of harmonic elimination equations. *IEEE Transactions on Power Electronics*. 1996;11(4):612–621.

[22] Al-Hitmi M, Ahmad S, Iqbal A, *et al.* Selective harmonic elimination in a wide modulation range using modified Newton–Raphson and pattern generation methods for a multilevel inverter. *Energies*. 2018;11(2):458.

[23] Liang TJ, O'Connell RM, and Hoft RG. Inverter harmonic reduction using Walsh function harmonic elimination method. *IEEE Transactions on Power Electronics*. 1997;12(6):971–982.

[24] Kato T. Sequential homotopy-based computation of multiple solutions for selected harmonic elimination in PWM inverters. *IEEE Transactions on Circuits and Systems I: Fundamental Theory and Applications*. 1999;46 (5):586–593.

[25] Ahmad S, Iqbal A, Ali M, *et al.* A fast convergent homotopy perturbation method for solving selective harmonics elimination PWM problem in multi level inverter. *IEEE Access*. 2021;9:113040–113051.

[26] Du Z, Tolbert LM, and Chiasson JN. Active harmonic elimination for multilevel converters. *IEEE Transactions on Power Electronics*. 2006;21(2):459–469.

[27] Yang K, Yuan Z, Yuan R, *et al.* A Groebner bases theory-based method for selective harmonic elimination. *IEEE Transactions on Power Electronics*. 2014;30(12):6581–6592.

[28] Zheng C and Zhang B. Application of Wu method to harmonic elimination techniques. *Proceedings of the CSEE*. 2005;25(15):40–45.

[29] Ozpineci B, Tolbert LM, and Chiasson JN. Harmonic optimization of multilevel converters using genetic algorithms. In: *2004 IEEE 35th Annual Power Electronics Specialists Conference* (IEEE Cat. No. 04CH37551). vol. 5. New York, NY: IEEE; 2004. p. 3911–3916.

[30] Kumari M, Ali M, Ahmad S, *et al.* Genetic algorithm based SHE-PWM for 1-ø and 3-ø voltage source inverters. In: *2019 International Conference on Power Electronics, Control and Automation (ICPECA)*. New York, NY: IEEE; 2019. p. 1–6.

[31] Kavousi A, Vahidi B, Salehi R, *et al.* Application of the bee algorithm for selective harmonic elimination strategy in multilevel inverters. *IEEE Transactions on Power Electronics.* 2011;27(4):1689–1696.

[32] Taghizadeh H and Hagh MT. Harmonic elimination of cascade multilevel inverters with nonequal DC sources using particle swarm optimization. *IEEE Transactions on Industrial Electronics.* 2010;57(11):3678–3684.

[33] Ahmad S and Iqbal A. Switching angles computations using PSO in selective harmonics minimization PWM. In: *Metaheuristic and Evolutionary Computation: Algorithms and Applications*. New York, NY: Springer; 2021. p. 437–461.

[34] Ahmad S, Iqbal A, Ashraf I, *et al.* Harmonics minimization in multilevel inverter by continuous mode ACO technique. In: *Innovations in Electrical and Electronic Engineering.* New York, NY: Springer; 2021. p. 95–104.

[35] Majed A, Salam Z, and Amjad AM. Harmonics elimination PWM based direct control for 23-level multilevel distribution STATCOM using differential evolution algorithm. *Electric Power Systems Research.* 2017; 152:48–60.

[36] Ahmad S, Al-Hitmi M, Iqbal A, *et al.* Low switching frequency modulation of a 3 × 3 matrix converter in UPFC application using differential evolution method. *International Transactions on Electrical Energy Systems.* 2020;30 (1):e12179.

[37] Ganie ZA, Bhat AH, and Ahmad S. Harmonics minimization in wide modulation range using improved differential evolution optimization technique and hardware validation. In: *2022 2nd Asian Conference on Innovation in Technology (ASIANCON)*. New York, NY: IEEE; 2022. p. 1–6.

[38] Patel HS and Hoft RG. Generalized techniques of harmonic elimination and voltage control in thyristor inverters: Part I—harmonic elimination. *IEEE Transactions on Industry Applications.* 1973;(3):310–317.

[39] Fei W, Ruan X, and Wu B. A generalized formulation of quarter-wave symmetry SHE-PWM problems for multilevel inverters. *IEEE Transactions on Power Electronics.* 2009;24(7):1758–1766.

[40] Dahidah M, Konstantinou G, Flourentzou N, *et al.* On comparing the symmetrical and non-symmetrical selective harmonic elimination pulse-width modulation technique for two-level three-phase voltage source converters. *IET Power Electronics.* 2010;3(6):829–842.

[41] Etesami M, Farokhnia N, and Fathi SH. Colonial competitive algorithm development toward harmonic minimization in multilevel inverters. *IEEE Transactions on Industrial Informatics.* 2015;11(2):459–466.

[42] Sun J and Grotstollen H. Solving nonlinear equations for selective harmonic eliminated PWM using predicted initial values. In: *Proceedings of the 1992 International Conference on Industrial Electronics, Control, Instrumentation, and Automation.* New York, NY: IEEE; 1992. p. 259–264.

[43] Barkati S, Baghli L, Berkouk EM, *et al.* Harmonic elimination in diode-clamped multilevel inverter using evolutionary algorithms. *Electric Power Systems Research.* 2008;78(10):1736–1746.

[44] Kumar J, Das B, and Agarwal P. Harmonic reduction technique for a cascade multilevel inverter. *International Journal of Recent Trends in Engineering*. 2009;1(3):181.

[45] Son GT, Chung YH, Baek ST, *et al.* Improved PD-PWM for minimizing harmonics of multilevel converter using gradient optimization. In: *2014 IEEE PES General Meeting—Conference & Exposition*. New York, NY: IEEE; 2014. p. 1–5.

[46] Janabi A, Wang B, and Czarkowski D. Generalized Chudnovsky algorithm for real-time PWM selective harmonic elimination/modulation: two-level VSI example. *IEEE Transactions on Power Electronics*. 2019;35(5):5437–5446.

[47] Chiasson JN, Tolbert LM, McKenzie KJ, *et al.* Control of a multilevel converter using resultant theory. *IEEE Transactions on Control Systems Technology*. 2003;11(3):345–354.

[48] Chiasson JN, Tolbert LM, Du Z, *et al.* The use of power sums to solve the harmonic elimination equations for multilevel converters. *EPE Journal*. 2005;15(1):19–27.

[49] Ni J, Wu L, Fan X, *et al.* Bioinspired intelligent algorithm and its applications for mobile robot control: a survey. *Computational Intelligence and Neuroscience*. 2016;2016: Article ID 3810903.

[50] Ab Wahab MN, Nefti-Meziani S, and Atyabi A. A comprehensive review of swarm optimization algorithms. *PLoS ONE*. 2015;10(5):e0122827.

[51] Holland JH. Genetic algorithms: computer programs that "evolve" in ways that resemble natural selection can solve complex problems even their creators do not fully understand. *Scientific American*. 2005;267:1992.

[52] Kennedy J and Eberhart R. Particle swarm optimization. In: *Proceedings of ICNN'95—International Conference on Neural Networks*, vol. 4. New York, NY: IEEE; 1995. p. 1942–1948.

[53] Karaboga D. (2005) An Idea Based on Honey Bee Swarm for Numerical Optimization. Technical Report-TR06, Erciyes University, Engineering Faculty, Department of Computer Engineering.

[54] Storn R and Price K. Differential evolution – a simple and efficient heuristic for global optimization over continuous spaces. *Journal of Global Optimization*. 1997;11(4):341–359.

[55] Atashpaz-Gargari E and Lucas C. Imperialist competitive algorithm: an algorithm for optimization inspired by imperialistic competition. In: *2007 IEEE Congress on Evolutionary Computation*. New York, NY: IEEE; 2007. p. 4661–4667.

[56] Yang XS. Firefly algorithms for multimodal optimization. In: *International Symposium on Stochastic Algorithms*. New York, NY: Springer; 2009. p. 169–178.

[57] Adeyemo I, Fakolujo O, and Adepoju G. Ant colony optimisation approach to selective harmonic elimination in multilevel Inverter. *IMPACT: International Journal of Research in Engineering & Technology*. 2015;3:31–42.

[58] Yang XS and Deb S. Cuckoo search via Lévy flights. In: *2009 World Congress on Nature & Biologically Inspired Computing (NaBIC)*. New York, NY: IEEE; 2009. p. 210–214.

[59] Yang XS. A new metaheuristic bat-inspired algorithm. In: *Nature Inspired Cooperative Strategies for Optimization (NICSO 2010)*. New York, NY: Springer; 2010. p. 65–74.

[60] Mehrabian AR and Lucas C. A novel numerical optimization algorithm inspired from weed colonization. *Ecological Informatics.* 2006;1(4): 355–366.

[61] Ahmad Ganie Z, Hamid Bhat A, and Ahmad S. A novel adaptive invasive weed optimization technique and least square regression for harmonics minimization in standalone PV applications. *International Journal of Circuit Theory and Applications.* 2022;54:1–23. Available from: https://onlinelibrary.wiley.com/doi/abs/10.1002/cta.3483.

Chapter 10

Comparison and overview of power converter control methods

Mohamed Bendaoud[1], Farid Oufqir[1], Farhad Ilahi Bakhsh[2], Khalid Chikh[1], Amine El Fathi[3] and Abdesslam Lokriti[1]

The control performance of the power electronics converter plays an important role in the modern electrical system. There are many advanced control methods that are intended to improve the control performance of these converters. In this chapter, a systematic review of the most advanced control methods has been developed. The advanced control methods are generally classified into two categories: linear controller and nonlinear controller. These are basically based on the characteristics of the controller and how to model the controlled installation. In addition, the performance comparison of these advanced control methods for different converters is performed, which can help engineers to choose the appropriate control methods for an individual converter.

10.1 Introduction

The use of power converters is becoming increasingly important. Current technological applications require a high level of precision and performance which depend not only on the power part but also on the control laws used to control the power switches of the converter. As shown in Figure 10.1, the control laws are divided into linear, nonlinear, and intelligent controls.

DC–DC converters are nonlinear systems, so linear controllers may be sufficient if the accuracy and performance requirements of the system are not too severe. They are often subject to significant disturbances, the operating point is no longer fixed at a nominal position and the control algorithms designed cannot,

[1]Science and Technology for the Engineer Laboratory (LaSTI), Sultan Moulay Slimane University, Morocco
[2]National Institute of Technology Srinagar, Hazratbal, Srinagar (J&K), India
[3]LRSDI Laboratory, Faculty of Sciences and Technology of Al Hoceima, Abdelmalek Essaadi University, Morocco

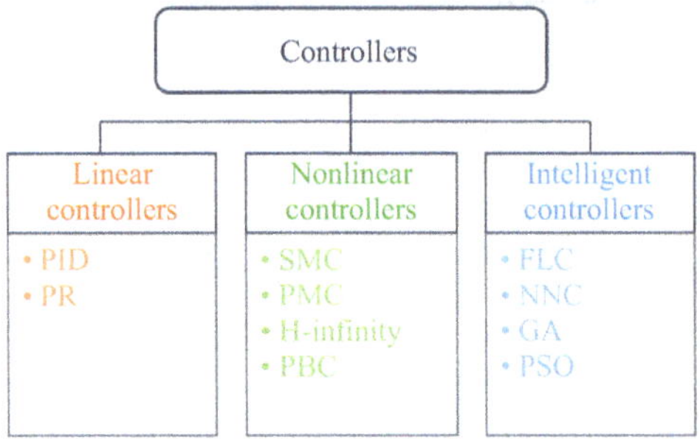

Figure 10.1 Classification of control methods

therefore, ensure robust behavior with respect to uncertainties on the parameters and their variations. This has led to an important interest in the synthesis of non-linear, robust control techniques capable of overcoming this problem. We can cite in this context adaptive control, sliding mode control (SMC), predictive control, robust control, etc.

Most nonlinear control methods require the availability of a mathematical model of the power converter. The precision of the model utilized will have a direct impact on the performance. To solve these problems, the employment of intelligent controllers can be an alternative. It includes fuzzy logic, neural network controllers, and metaheuristic methods.

The objective of this chapter is to provide a comparison of the control methods for DC–DC converters and DC–AC inverters. In this chapter, nonlinear control approaches will be presented and discussed in Section 10.2. Intelligent control approaches will be presented and discussed in Section 10.3. Then, Section 10.4 will present a comparison of these controllers in order to determine the appropriate control approach for a given application. Finally, a conclusion is drawn in Section 10.5.

10.2 Nonlinear controllers for power converters

In this section, nonlinear controllers used for the control of power electronics converters are discussed.

10.2.1 Sliding mode

The SMC approach has gained recognition as one of the most widely used and effective control methods in power converters during the past decades. The SMC is a variable structure control where the change of structure is generally carried out by switching at the level of the control device. The objective of this method is to constrain the trajectory of the system, using a discontinuous command, to evolve and maintain itself, beyond a finite time, on a surface, called "sliding surface" $S(x)$, where the resulting behavior is the desired dynamics. In other words, the objective is to arrive at the reference state, once the state of the system reaches the sliding surface (Figure 10.2), the system becomes in the sliding regime. Its dynamic is then

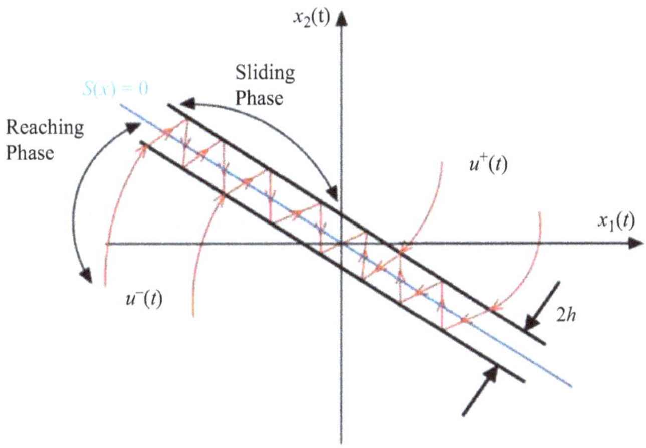

Figure 10.2 The behavior of trajectories under a SMC

insensitive to external disturbances as long as the conditions of the sliding regime are ensured [1].

This control approach has several advantages such as robustness to disturbances, precision, stability, and simplicity. Despite these attractive benefits, the SMC suffers from chattering in state variables which leads to low control accuracy and losses in power converters.

As shown in Figure 10.3, the SMC approach can be classified into the sliding surface, the control law, the type of modulation, and the chattering reduction methods.

- Sliding surface

  The sliding surface is defined with respect to the order of the model of the system to be controlled. Indeed, in the simple case, it is defined with a linear function of the error [2]:

$$S = \alpha^t e(t) \tag{10.1}$$

where α is a vector that contains the sliding coefficients.

And $e(t) = x(t) - x_d(t) = (e(t)\ \dot{e}(t)\ \ldots\ e(t)^{n-1})$ is the deviation (error) of the variable to be adjusted, with $x_d(t)$ being the desired value. In this case, it is a conventional SMC based on a PD sliding surface. An integral action can be added to obtain a sliding surface-based PID:

$$S = \alpha_1 e(t) + \alpha_2 \dot{e}(t) + \alpha_3 \int e(t)dt \tag{10.2}$$

To improve the robustness of the system to be controlled as well as the elimination of the reaching mode, a nonlinear sliding surface has been proposed by several authors based on ellipsoidal, lemniscate, or parabolic [3].

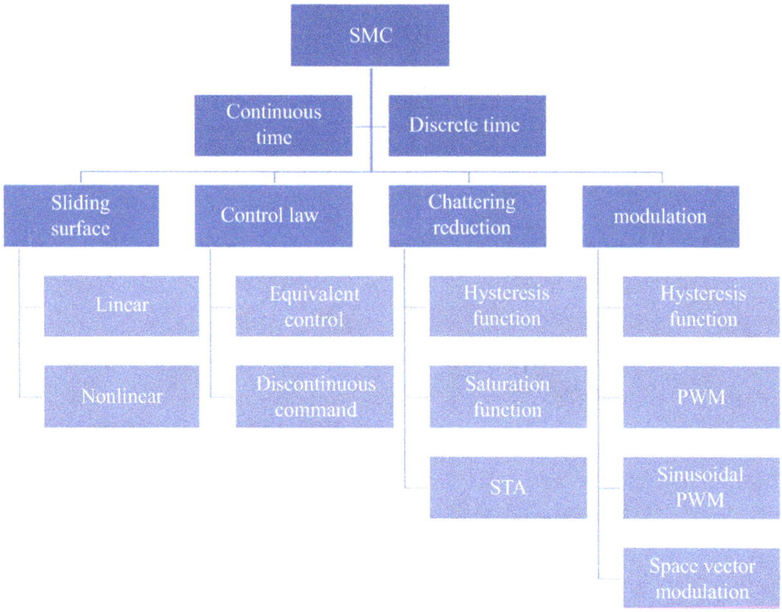

Figure 10.3 SMC classification

- Control law

Once the sliding surface has been selected, it remains to determine the control law necessary to attract the variable to be regulated on the surface, and then towards its point of equilibrium, while maintaining the condition of existence sliding mode. The control law has two parts:

$$u = u_{eq} + u_n \tag{10.3}$$

u_{eq} corresponds to the continuous part and u_n corresponds to the discontinuous part of the command.

The equivalent command proposed by FILIPOV and UTKIN makes it possible to maintain the variable to be regulated on the sliding surface $S = 0$. The equivalent command is calculated [4], considering that the derivative of the surface is zero $\dot{S} = 0$.

This command can be interpreted otherwise as being an average value that the command takes during a rapid switching between the values u^+ and u^-.

The discontinuous command is determined by the values taken by the sliding surface S in the following way:

$$u = -K.\,sign(S) \tag{10.4}$$

where K is a positive constant representing the gain of the discontinuous control.

These two commands can be used together, or only one component can be used, either the continuous command or the discontinuous command.

Using the sign function results in chattering, which is not desired in power converters. To reduce the effect of this phenomenon, several methods have been proposed in the literature.

- Chattering reduction

The first solution consists in replacing the sign function by a hysteresis function characterized by one or two thresholds. The principle of this method is shown in Figure 10.4.

Although this solution is easy to implement, it does not allow fixing the switching frequency.

The second solution consists in using the saturation function defined by:

$$\text{sat(S)} = \begin{cases} -1 \text{ if } S > \phi \\ \dfrac{S}{\phi} \text{ if } |S| < -\phi \\ 1 \text{ if } S < -\phi \end{cases} \tag{10.5}$$

This makes the control signal continuous and consequently, the switching frequency becomes constant [5,6]. The principle of this method is shown in Figure 10.5.

The super convolution algorithm (STA) adapted to a relative first-degree system has been proposed as a solution to reduce the chattering effect. In this case, the control law is defined by:

$$u(t) = -\alpha|S|^{0.5}sign(S) - \beta\int sign(S)dt \tag{10.6}$$

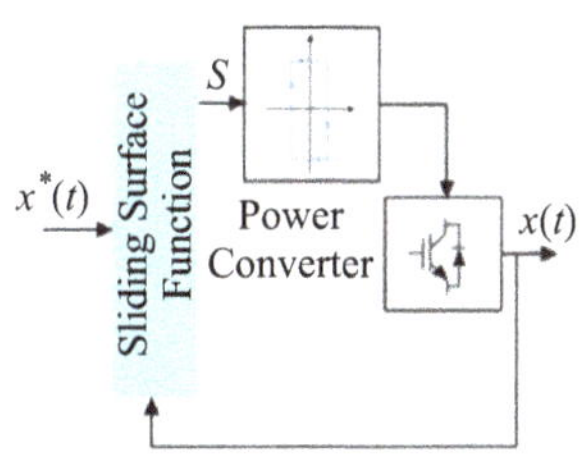

Figure 10.4 Block diagram of the SMC with hysteresis function

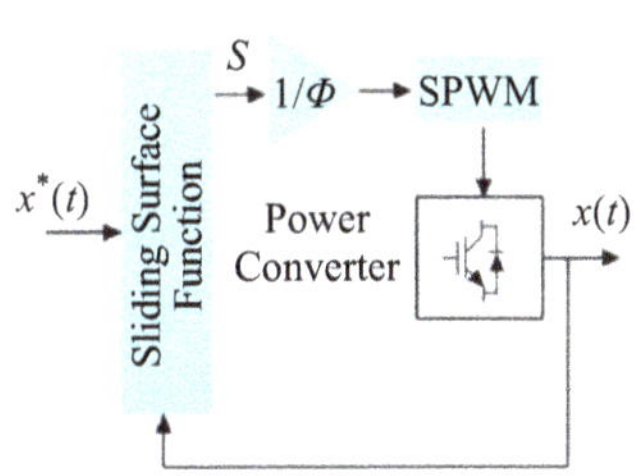

Figure 10.5 Block diagram of the SMC with saturation function

where α and β are positive constants that are utilized to adjust the STA for the desired performance. The principle of this method is shown in Figure 10.6.

10.2.1.1 The SMC for DC–DC converters

The control structure of a DC–DC converter by SMC is given in Figure 10.7.

In the literature, different SM controllers have been proposed for DC–DC converters. The implementation of a conventional SMC with hysteresis modulation has been presented in [8] for a buck converter, and in [11] for a boost converter. To improve control performance and to reduce the effect of chattering, Tan *et al.* implemented the SMC based on sliding surface PID with a PWM modulation in the buck [9], boost [12], and buck–boost [12,14] converters. In [10], Komurcugil proposed the use of a nonlinear sliding surface with hysteresis modulation to control a buck converter. Furthermore, the same method was applied in a boost converter with PWM modulation in [13].

In [16], Malesani presented the control of the Cuk converter by conventional SMC with hysteresis modulation (HM). Then, the command was extended to other more complex converters such as SEPIC [15] and ZETA [17]. Table 10.1 summarizes the performances obtained for the different SMC applied to the DC–DC converters.

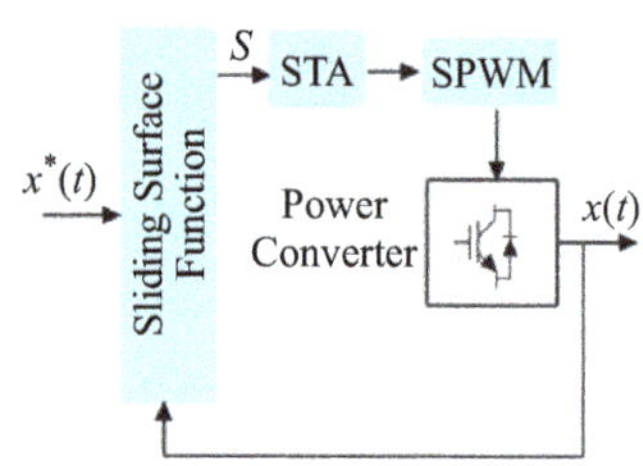

Figure 10.6 Block diagram of the SMC with STA

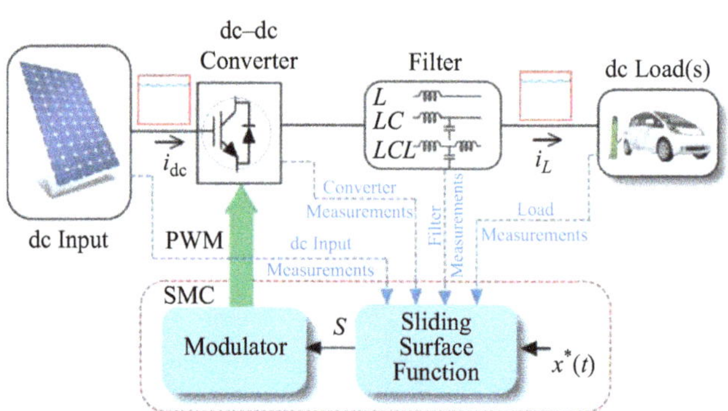

Figure 10.7 The control structure of a DC–DC converter by SMC [7]

Table 10.1 Performances obtained by SM controllers applied to DC–DC converters

Converter topology	SMC category	Modulation	Switching frequency	Voltage tracking	Robustness under variation of		References
					Input voltage	Load	
Buck	Conventional SMC	HM	Variable	NT	+	+	[8]
	PID SMC	PWM	200 kHz	NT	+	++	[9]
	Terminal SMC	HM	15.625 kHz	+	++	++	[10]
Boost	Conventional SMC	HM	Variable	NT	NT	++	[11]
	PID SMC	PWM	200 kHz	NT	NT	+	[12]
	Terminal SMC	PWM	20 kHz	++	+	+	[13]
Buck-boost	PID SMC	PWM	9.25 kHz	++	+	+	[14]
SEPIC	Conventional SMC	HM	Variable	NT	++	++	[15]
Cuk	Conventional SMC	HM	Variable	++	NT	++	[16]
Zeta	PI-SMC	HM	20 kHz	+	+	+	[17]
Flyback	PID-SMC	PWM	40 kHz	+	+	+	[18]
Forward	PID-SMC	PWM	5 kHz	NT	NT	++	[19]
Interleaved boost converter	Conventional SMC	HM	40 kHz	++	++	+	[20]

Good, + very good, ++; NT: not tested; NS, not specified.

10.2.1.2 The SMC for DC–AC inverters

Figure 10.8 shows the structure of SMC controlling an inverter that operates in grid-connected or standalone mode.

In the literature, different SM controllers have been applied to DC–AC inverters. Table 10.2 shows the performances obtained for the different SMC applied in the inverters operating in grid-connected or in standalone mode.

10.2.2 Model predictive control

Model predictive control (MPC) is another type of control that is widely used in the field of power converters. This control method is efficient in terms of speed, accuracy, and stability in power converter applications.

Figure 10.9 shows the basic diagram of the MPC strategy which contains a prediction block to predict the evolution of the output y_p. The prediction is made by an internal model of the system over a limited period called the prediction horizon. The second block uses the references and the predicted variables to determine the control signal making the reference rejoin according to a predefined trajectory (reference trajectory) on the output of the process by minimizing a cost function.

The cost function expresses the weighted sum of the differences or errors between the variables to be controlled and the references in addition to the secondary terms which include the constraints, the security measures, etc. In general, the cost function has the following form:

$$J = \sum_{j=N_1}^{N_2} \left[y_p\left(k + \frac{j}{k}\right) - y_{ref}\left(k + \frac{j}{k}\right) \right]^2 + \sum_{j=1}^{N_u} \lambda [\Delta u(k + j - 1)]^2 \quad (10.7)$$

where $y_p\left(k + \frac{j}{k}\right)$ is the predicted outputs, $j = 0 \ldots N$ (N is the prediction horizon), $y_{ref}\left(k + \frac{j}{k}\right)$ is the set point applied at instant $k + \frac{j}{k}$, $\Delta u(k + j - 1)$ is the command increment at instant $j - 1$, $N_p = N_2 - N_1$ is the prediction horizon, N_u is the control horizon, and λ is the weighting factor.

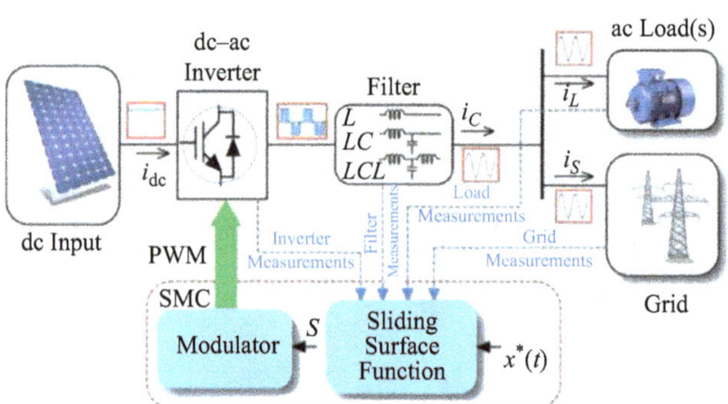

Figure 10.8 Control structure of a DC–AC inverter by SMC [7]

Table 10.2 Performances obtained for the different SMCs applied to inverters

Application	Inverter topology	SMC category	Modulation	Switching frequency	Sim /Exp	THD of Vo	THD of Io	Control objective	Response to sudden changes
Stand-alone	Single-phase [21]	Conventional SMC	HM	Variable	Exp	3.4%	NS	Voltage regulation for different load types	++
	Single-phase [22]	Conventional SMC	PWM	15 kHz	Exp	1.1%	NS	Voltage regulation for different load types	+
	Four-leg Inverter (3-ph) [23]	Conventional SMC	SPWM	15 kHz	Exp	1.7% (NL) 0.4 (L)	NS	Voltage regulation for different load types	+
Grid-conn-ected	Single-phase [24]	Multi-resonant SMC	PWM	15 kHz	Exp	NS	0.76%	-Current regulation -THD minimization	++
	H-bridge (3-ph) [25]	SM observer-based Control	PWM	6 kHz	Exp	NS	1.5%	-Power regulation -Control robustness under grid voltage unbalance and distortion, and system parameter deviation	++
	3L-NPC [26]	Nonlinear SMC	SPWM	2 kHz	Exp	NS	1.8%	-DC voltage regulation -Current regulation -THD minimization	++
	1L-Z-source [27]	Hyper-plane SMC	SPWM	18 kHz	Exp	NS	1.1%	-DC voltage regulation -Current regulation -THD minimization	+

Good, +; very good, ++; NT, not tested; NS, not specified; L, linear; NL, nonlinear.

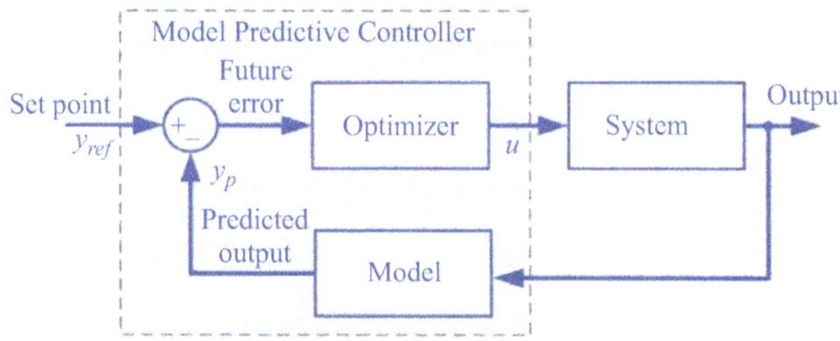

Figure 10.9 Principle block diagram of the MPC

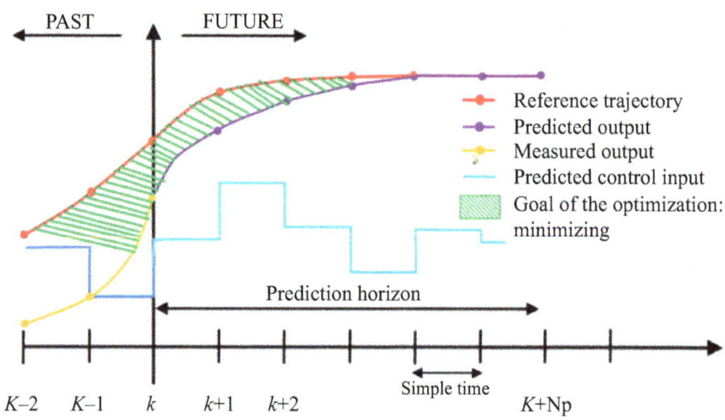

Figure 10.10 Time evolution of finite horizon prediction

Figure 10.10 shows how this strategy works over a finite horizon. In general, the predictive control law is obtained from the following methodology [28]:

- Predict the future outputs of the process over the defined prediction horizon, using the prediction model.
- The future control signals $u\left(k+\frac{j}{k}\right)$ are calculated by optimizing a cost function so that the output y of the process is as close as possible to the reference trajectory y_{ref}.
- Finally, only the first element of the vector of the control signal u is applied to the system during the next instant.

In the literature, different strategies based on predictive control have been proposed for controlling power converters. Nevertheless, all these techniques use the same control philosophy and the principle of operation remains the same. The classification of predictive control based on a model is shown in Figure 10.11. According to this figure, the MPC can be classified into two large families [29].

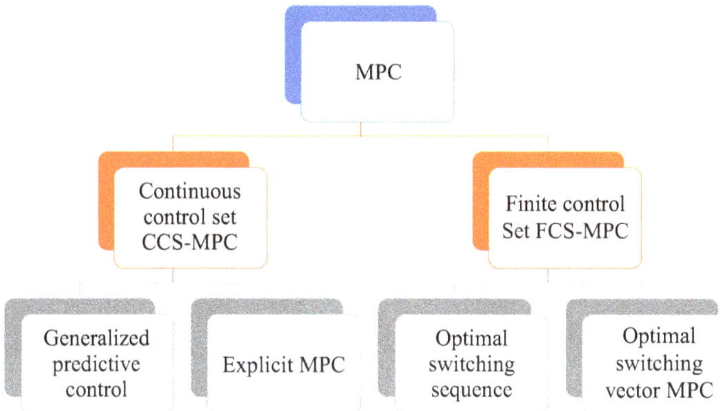

Figure 10.11 Classification of MPC methods applied to power converters

In the first category (continuous control set CCS-MPC), the control variables are continuous, and the optimization problem can be solved online and real-time, generally using numerical programming. In this technique, a modulator is needed to generate the switching states, which leads to a fixed switching frequency. This command category is also divided into two classes: explicit MPC (EMPC) used for nonlinear systems and generalized predictive control (GPC) used for linear systems. EMPC performs off-line optimization and can therefore be implemented on low-cost embedded systems.

The second category (finite control set FCS-MPC) takes advantage of the discrete nature of the converters to solve the optimization problem and does not require a modulator. The principle consists of predicting the output variables that we want to control for each switching state, then the suitable switching state that minimizes the cost function is chosen. With this control strategy, the complexity of the constrained optimization problem can be significantly reduced. This command category is also divided into two classes: optimal switching sequence MPC (OSS-MPC) and optimal switching vector MPC (OSV-MPC). The switching vectors (SVs) are used by the OSV-MPC to define the feasible control set. The OSS-MPC establishes a set of switching sequences, each of them is composed of a specific number of SVs. Table 10.3 presents a detailed comparison between these MPC methods.

10.2.2.1 The MPC for DC–DC converters

In the literature, different MP controllers have been applied to DC–DC converters. We start with the EMPC, which has been used in several power converters. The off-line implementation of the EMPC has been presented in [30] for a boost converter feeding constant power load. To control a buck converter with variable load, Chen *et al.* [31] proposed to linearize the model of the converter. Using this model, the EMPC strategy can be applied to the converter in order to obtain off-line control

Table 10.3 Comparison between MPC methods

	Continuous control set		**Finite control set**	
	GPC	**EMPC**	**OSS-MPC**	**OSV-MPC**
Optimization	Online	Offline	Online	Online
Constraints	Can be allowed but increase the computational cost	Allowed	Allowed	Allowed
Prediction Horizon	Long	Long	Mostly short	Mostly short
Modulator	SVM or PWM	SVM or PWM	Not required	Not required
Switching frequency	Fixed	Fixed	Fixed	Variable
Algorithm complexity	Complex	Complex	Intuitive	Intuitive

laws independent of the load, which makes it possible to avoid the degradation of the performance of the control. The implementation of EMPC in an FPGA device has been presented in [32] to control a buck–boost converter.

FCS-MPC strategies have also been widely used to control DC–DC converters. In [33], an FCS-MPC has been proposed to control a single-input dual output fly-back converter. The major drawback of this control method is the variation in switching frequency. To solve this problem, Guler *et al.* [34] have proposed an auto-tuning weighting factor that makes it possible to maintain the switching frequency despite the variation of the internal parameters of the SEPIC converter. In [35], an FCS-MPC was applied to a boost converter in order to extract maximum power from a PV. Table 10.4 summarizes the performances obtained for the different MPC applied to the DC–DC converters.

10.2.2.2 The MPC for DC–AC inverters

In the literature, different MP controllers have been proposed for DC–AC inverters. Table 10.5 shows the performances obtained for the different MPC applied in the inverters operating in grid-connected or in standalone mode. The table also shows the objectives of the command as well as the THD of the current and the voltage obtained.

FCS-MPC has been widely used to control inverters and especially multi-level inverters because of the finite number of switching states they present. For example, for a single-phase inverter, only four states are evaluated by the cost function. On the other hand, for CHB inverter, there are a high number of possible switching combinations so the optimization of the cost function becomes excessively high.

CCS-MPC has also been applied in a wide range of converters such as three-phase two-level converters, flying capacitor converter and neutral point-clamped (NPC) inverters.

Table 10.4 Performances obtained by MP controllers applied to DC–DC converters

Converter topology	MPC category	Np	Modulator	Switching frequency (kHz)	Voltage tracking	Robustness under variation of		Reference
						Input voltage	Load	
Boost	EMPC	2	NS	50	++	++	NT	[30]
	FCS-MPCS	NS	Without	100	++	++	NT	[35]
Buck	EMPC	10	NS	250	NT	NT	++	[31]
Buck–boost	EMPC	4	PWM	NS	NT	++	NT	[32]
	AMPC	100	PWM	25	++	NT	+	[36]
SEPIC	FCS-MPC	NS	Without	10	+	+	+	[34]
Cuk	FCS-MPC	2	Without	15.68	++	NT	++	[37]
Dual-output flyback converter	FCS-MPC	1	Without	Variable	++	NT	++	[33]
Forward	EMPC	NS	PWM	NS	++	NT	+	[38]
DAB	FCS-MPC	21	Without	20	+	NT	++	[39]

Good, +; very good, ++; NT, not tested; NS: not specified.

Table 10.5 Performances obtained by MP controllers applied to DC–AC inverters

	Inverter topology	MPC category	Np	Switching frequency (kHz)	THD of V_o (%)	THD of I_o (%)	Control objective	Response for sudden changes	References
Stand-alone	Single phase	FCS	2	Variable	2.23 (L) 2.87 (NL)	NS	Voltage regulation for different type of load	+	[40]
	FCS(3-ph)	FCS	1	Variable	NS	NS	Voltage regulation for different type of load	+	[41]
		CCS	1	40	NS	NS	Voltage regulation for different type of load	++	[42]
	H-bridge (3-ph)	CCS	NS	NS	4.6 (NL) 2.8 (L)	NS	Voltage regulation for different type of load	+	[43]
		FCS	1	Variable	1.7 (NL) 0.4 (L)	NS	Voltage regulation for different type of load	+	[44]
Grid-connected	NPC(1-ph)	FCS	1	5	NS	5	-Current regulation -THD minimization	NT	[45]
	H-bridge (3-ph)	CCS	8	10	NS	4.2	-Power regulation -Control robustness under grid voltage unbalance and distortion, and system parameter deviation	++	[46]
		FCS	NS	Variable	NS	2.7	-Power regulation -THD minimization	NT	[47]
	3L-NPC	FCS	1	10.5	NS	2.7	-DC voltage regulation -Current regulation -THD minimization	NT	[48]
	CHB(3-ph)	FCS	1	Variable	NS	1.97	-DC voltage regulation -Current regulation -THD minimization	NT	[49]
	CHB(1-ph)	FCS	1	Variable	NS	1.32	-DC voltage regulation -Current regulation -THD minimization	++	[50]

Good, +; very good, ++; NT, note tested; NS, not specified; L, linear; NL, nonlinear.

10.3 Intelligent controllers for power converter

In this section, intelligent controllers used for the control of power electronics converters are discussed.

10.3.1 Fuzzy logic controller (FLC)

Most nonlinear control methods require the availability of a mathematical model of the system. The accuracy of the model used will have a direct impact on the performance achieved. To solve these problems, the use of controllers based on human expertise can be an alternative. One of these methods is fuzzy logic control that compensates the uncertainties as well as the nonlinearities neglected during the mathematical modeling of the system.

Fuzzy systems make it possible to efficiently exploit and manipulate linguistic information emanating from the human expert, thanks to a large theoretical arsenal [51]. In addition, the system implemented can be easily integrated into a control or identification loop. The basic structure of a fuzzy system is divided into three main parts as shown in Figure 10.12.

- Fuzzification

  The input x varies in a domain called the universe of discourse X, split into a finite number of fuzzy sets such that there is a dominant situation in each area. In order to facilitate the numerical processing and the use of these sets, they are described by membership functions (MFs). They accept as an argument the position of x in the universe of discourse, and as output the degree to which x belongs to the situation described by the function. Figure 10.13 gives some examples of membership functions.

  The term linguistic variable designates a variable defined in a universe of discourse and taking a certain value like small, large, etc.

  Fuzzification consists in defining membership functions for the different linguistic variables. It is a projection of the physical variable onto the fuzzy sets characterizing this variable.

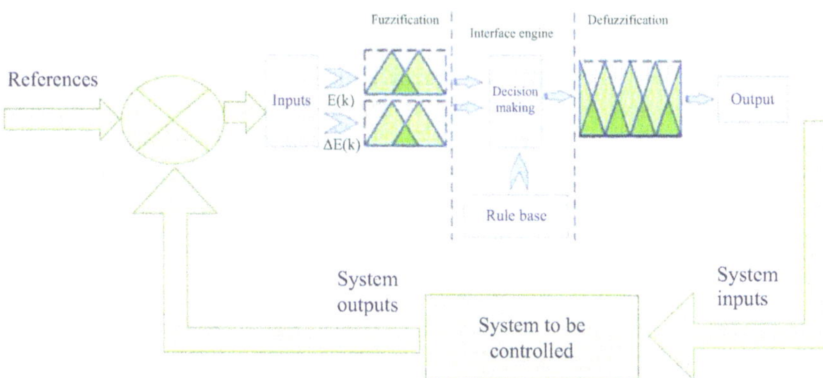

Figure 10.12 FLC architecture

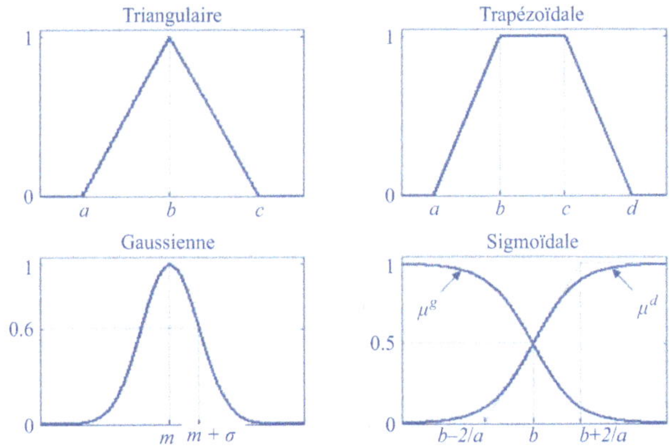

Figure 10.13 Membership functions

- Fuzzy interference

  The values of linguistic variables, defined by MFs, are linked together by rules, in order to deduce results from them. A fuzzy rule of inference or a fuzzy relationship is often expressed by the conditional logical structure "If-then." Compared to Sugeno method, the Mamdani approach is more often used for the inference process.
- Defuzzification

  At the end of the inference, the output fuzzy set is determined but it is not directly used to control the system. It is necessary to move from the "fuzzy world" to the "real world," this is achieved by defuzzification. There are a number of methods that make it possible to find an output value i.e. the maximum method, the method of the barycenter of maxima and the method of the center of gravity.
- Type-2 fuzzy logic

  As an expansion of the concept of ordinary fuzzy sets, also known as type-1 fuzzy sets, Zadeh [52] established the concept of type-2 fuzzy sets. The ability of type-2 fuzzy logic to take linguistic and numerical uncertainties into account is its main advantage over type-1 fuzzy logic. The type-2 fuzzy logic system defines the system uncertainty across a bounded region in the membership function as upper and lower MF [53]. The fundamental structure of a type-2 fuzzy system is shown in Figure 10.14. It resembles the structure of a type-1 fuzzy system. The output processor is composed of a defuzzifier and a type reducer which transforms a fuzzified type-2 fuzzy set to a type-1 fuzzy set.

10.3.1.1 The FLC for DC–DC converters

In the literature, different FL controllers have been proposed for DC–DC converters. In reality, there are a few research works on the control of DC–DC converters by FLC [54,55]. Most of the papers use FLC to control the converters in order to extract the maximum power from the photo-voltaic (PV). Table 10.6

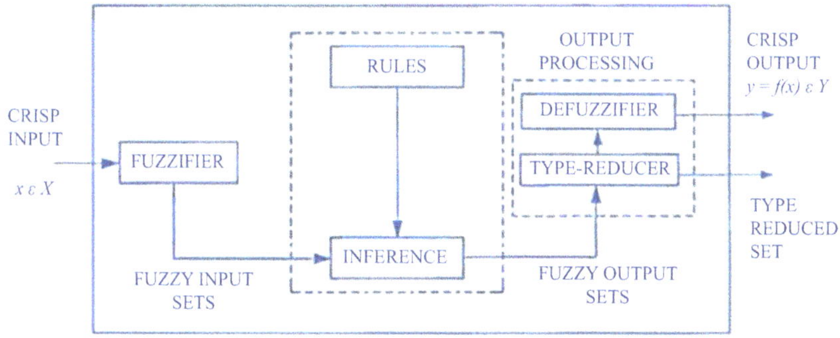

Figure 10.14 The fundamental structure of type-2 fuzzy system

Table 10.6 Performances obtained by FL controllers applied to DC–DC converters

Converter topology	FLC category	Number of rules	Switching frequency (kHz)	Voltage tracking	Robustness under variation of		Reference
					Input voltage	Load	
Buck	-Type-1 -Mamdani inference	49	31.373	++	NT	+	[54]
	Type-2	25	20	++	++	++	[56]
Boost	-Type-1 -Mamdani inference	49	31.373	++	NT	+	[54]
	Type-2	25	20	++	++	++	[56]
Buck–boost	-Type-1 -Mamdani inference	25	50	NT	NT	++	[55]
SEPIC	-Type-1 -Mamdani inference	25	10	++	++	NT	[57]
Cuk	-Type-1 -Mamdani inference	9	10	+	+	+	[58]
Flyback converter	-Type-1 -Sugeno inference	25	100	+	NT	NT	[59]
LLC resonant	-Type-1 -Mamdani inference	25	NS	NT	++	++	[60]
DAB	-Type-1 -Mamdani inference	9	10	NT	NT	+	[61]

Good, +; very good, ++; NT, not tested; NS, not specified.

summarizes the performances obtained for the different FLC applied in the DC–DC converters.

10.3.1.2 The FLC for DC–AC inverters

In the literature, different FL controllers have been proposed for DC–AC inverters. Table 10.7 shows the performances obtained for the different FLC applied to the inverters operating in grid connected or in standalone mode. The table also shows the objectives of the command as well as the THD of the current and the voltage obtained.

10.3.2 Artificial neural network

An artificial neural network (ANN) is a computer algorithm whose design is inspired by the biological neural networks of the human brain. The design of ANNs is based on the association in a more or less complex process of a set of simple mathematical operations called formal neurons [67]. A formal neuron is shown in Figure 10.15. Its output y is obtained in two stages. The weighted sum of the weights w_i and the input vector x_i, plus a bias b, is first calculated. Subsequently, the obtained result v passes through an activation function $f(.)$ which is generally nonlinear such as logistic sigmoid or hyperbolic tangent, to ensure the universal approximation property. Finally, the expression of the output y of the formal neuron can be written as follows:

$$y = f(v) = f\left(\sum_{i=1}^{n} x_i w_i + b\right) \tag{10.8}$$

The variables of this function are usually called "inputs" of the neuron.

Given the growing interest in ANNs, several structures have been proposed. They can be classified according to two criteria. The first criterion concerns the architecture adopted and the second concerns the type of learning chosen. There are mainly two architectures of ANN:

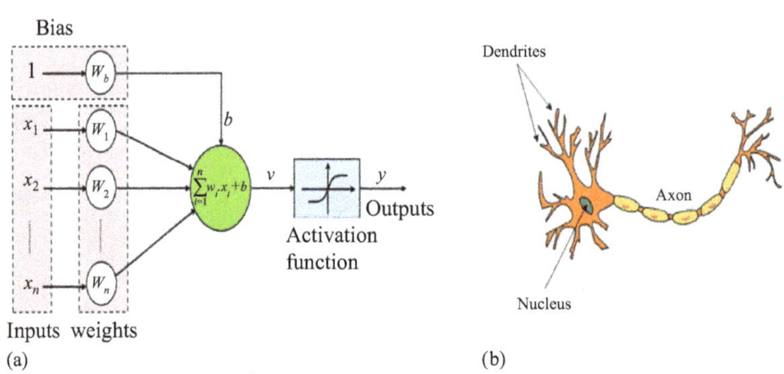

Figure 10.15 (a) Model of artificial neuron and (b) structure of biological neuron

Table 10.7 Performances obtained for different FLC applied to DC–AC inverters

	Inverter topology	FLC category	Number of rules	Switching frequency	Sim/Exp	THD of V_o (%)	THD of I_o (%)	Control objective	Response for sudden changes
Standalone	H-bridge (3-ph) [62]	Type-1	16	2 kHz	Sim	2 (L) 10 (NL)	NS	Voltage regulation for different type of load	+
	Quasi-z-source (1-ph) [63]	Type-1	25	10 kHz	Sim	1.3%	NS	MPPT	+
	CHB [64]	-Type-1 -Mamdani inference	45	14 kHz	Sim	1.3%	1.3%	MPPT	++
Grid-connected	NPC (3-ph) [65]	Type-1	49	10 kHz	Exp	NS	3.7%	MPPT	+
	H-bridge (3-ph) [66]	Type-1	42	NS	Sim	2.48%	4.6%	Current regulation	+
	CHB (3-ph) [64]	-Type-1 -Mamdani inference	45	14 kHz	Sim	0.5%	3.1%	MPPT	++

Good, +; very good, ++; NT, not tested; NS, not specified; L, linear; NL, nonlinear.

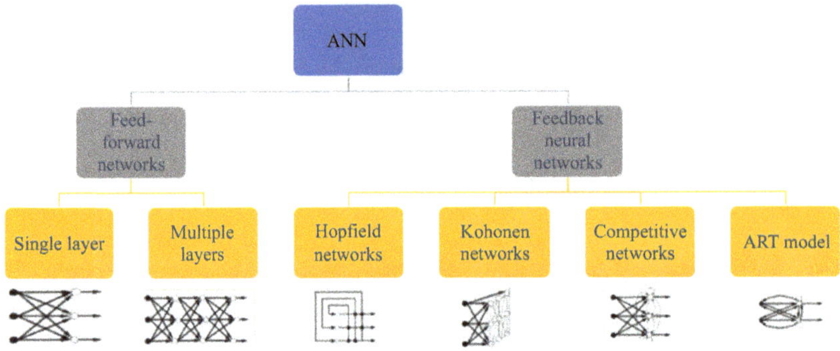

Figure 10.16 Classification of ANNs

Feed-forward networks: This architecture is characterized by the fact that the information propagates only from the input layer to the output layer, through any intermediate layers, with no possible backtracking. As shown in Figure 10.16, neurons are organized in the form of a single layer or multiple layers (an input layer, as many hidden layers as necessary and an output layer) [68].

Recurrent or feedback neural networks: Also called looped networks, these networks have a similar structure to non-looped networks. However, in this case, the connection of neurons is possible between elements of the same layer and also with previous layers. The information is thus conveyed both in forward propagation and in back propagation. This property makes these ANNs more powerful, since their operation is sequential and adopts a dynamic behavior. As seen in Figure 10.16, the four categories of the feedback neural network are the adaptive resonance theory model, the Hopfield networks, the Kohonen networks, and the competitive networks [68].

Once the ANN architecture is adopted, a learning method must be chosen to update the ANN weights. The learning results in a modification of the value of the weights connecting the neurons of the network. The learning process offers the possibility for the ANN to learn and then improve its performance.

There are mainly three learning methods i.e. supervised learning, unsupervised learning, and reinforcement learning. Their principles are presented below:

Supervised learning: It consists of adjusting the weights of the ANN in such a way as to minimize the error between the output provided by the ANN and the desired output produced by a supervisor excited by the same input. Supervised learning is illustrated conceptually in Figure 10.17.

Unsupervised learning: In the case of unsupervised learning, the adaptation of the weights depends only on the internal criteria of the network. The ANN analyzes all the data, and a compliance criterion tells it how close it is to the desired result. The weights of the ANN are then adapted to increase the precision of the algorithm. The principle of this method is shown in Figure 10.18.

Reinforcement learning: In this case, the ANN gradually learns from its mistakes. It is rewarded by the environment in which it interacts when the result is

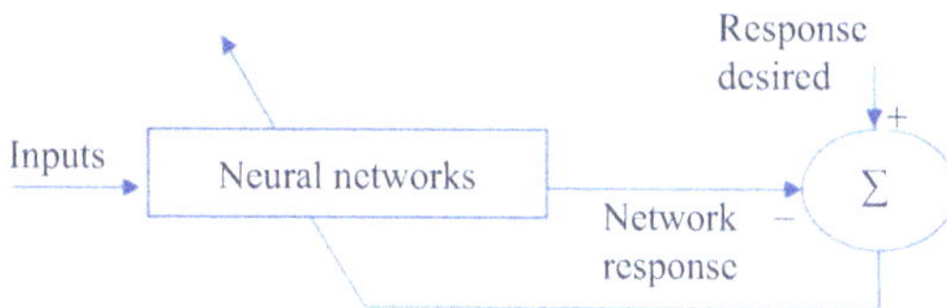

Figure 10.17 Principle of supervised learning

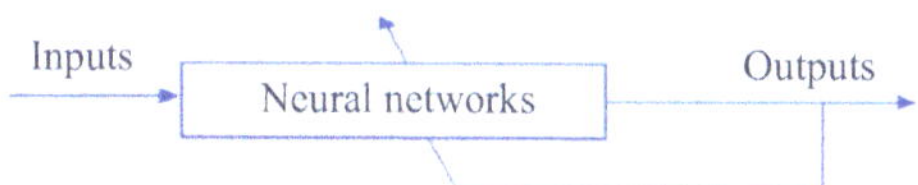

Figure 10.18 Principle of unsupervised learning

satisfactory, while it is penalized in the opposite case. In this way, the ANN seeks to maximize the sum of the rewards over time. The choice of an architecture or a learning method depends not only on the nature of the application but also on the processing capabilities of the processor in which the ANN will be implemented.

10.3.2.1 The ANN for DC–DC converters

In the literature, different ANN controllers have been proposed for DC–DC converters. In [69], a feed-forward neural controller was designed to regulate the output voltage of the buck converter. The ANN has been trained to implement an approximate dynamic programming-based control through several training experiences. In [70], a feed-forward neural controller was proposed to control a boost converter for DC microgrid applications. The authors used data from a model predictive controller to train ANN controller. In the article [71], a recurrent neural controller is proposed to extract the maximum power from the PV and also to limit the loss of shading power. In [72], a Hopfield neural network is proposed to determine the optimal parameters of the PI controller for the SEPIC converter. In [73], an adaptive neuro-fuzzy inference system (ANFIS) was designed to control the output voltage of the flyback converter. ANFIS uses NN to adjust MF parameters while adjusting fuzzy rules. Table 10.8 summarizes the performances obtained for the different NN controllers applied to the DC–DC converters.

10.3.2.2 The NNC for DC–AC inverters

Table 10.9 shows the performances obtained for the different NN controllers applied to inverters operating in grid-connected or in standalone mode. The table also shows the objectives of the command as well as the THD of the current and the voltage obtained.

In [78], the authors used a neural controller to control the output voltage of a three-phase inverter for different loads while minimizing the THD. To train ANN controller, the authors used data from a model predictive controller. Similarly, authors in [79] proposed a feed-forward neural controller for a three-phase neutral

Table 10.8 Performances obtained by NN controllers applied to DC–DC converters

Converter topology	ANN type	No. of neurons	NN objective	Switching frequency	Voltage tracking	Robustness under variation of		Reference
						Input voltage	Load	
Buck	FFNN	NS	Voltage regulation	10 kHz	++	++	+	[69]
Boost	FFNN	15	Voltage regulation	20 kHz	++	NT	++	[70]
	RNN	31	MPPT	NS	NT	++	NT	[71]
Buck–boost	FFNN	6	Voltage regulation	25 kHz	++	++	NT	[74]
SEPIC	HNN	NS	Voltage regulation	50 kHz	+	+	+	[72]
Cuk	FFNN	8	Voltage regulation	20 kHz	+	NT	+	[75]
Flyback	FFNN	5	Voltage regulation	PWM	+	NT	NT	[76]
DAB	ANFIS	NS	Voltage regulation	100 kHz	++	++	++	[73]
	FFNN	50	Voltage regulation	50 kHz	+	+	NT	[77]

Good, +; very good, ++; NT, not tested; NS, not specified.

Table 10.9 Performances obtained for different NNC applied to DC–AC inverters

Application	Inverter topology	ANN type	No. of neurons	Switching Frequency	Sim/Exp	THD of V_o	THD of I_o	Control objective	Response for sudden changes
Standalone	H-bridge (3-ph)	FFNN [78]	NS	NS	Sim	1.6% (L) 1.87% (NL)	NS	Voltage regulation for different type of load	++
	NPC (3-ph)	FFNN [79]	64	5.7 kHz	Exp	1.6%	NS	-DC voltage regulation -Load voltage regulation	++
	Flying capacitor (3-ph)	FFNN [81]	NS	Variable	Exp	NS	2.82%	-DC voltage regulation -Load voltage regulation	++
Grid-connected	NPC (3-ph)	FFNN [82]	10	10 kHz	Sim	0.13%	3.49%	-DC voltage regulation -Current regulation -THD minimization	++
	H-bridge (3-ph)	FFNN [80]	26	10 kHz	Sim	NS	0.17%	- MPPT -THD minimization	+
	H-bridge (1-ph)	FFNN [83]	18	6 kHz	Exp	NS	NS	Power regulation	++

Good, +; very good, ++; NT, not tested; NS, not specified; L, linear; NL, nonlinear.

point-clamped converter. The proposed method uses the FS-MPC to generate a dataset for use in the training phase. However, the switching mechanism is similar to FS-MPC. Therefore, the switching frequency will be variable. In [80], a feed-forward neural controller is proposed to control the power injected into the grid by a three-phase inverter and to minimize the THD.

10.3.3 Metaheuristic optimization

The relationships between automatic control and optimization are very strong because optimization techniques are often at the core of automatic control methodologies. Indeed, optimization has traditionally provided effective tools for identifying system models, calculating control laws, analyzing the stability and robustness of systems, etc.

The optimization aims to find values for a parameter that will optimize an objective. The classical approach or the local heuristic search can be used to search for a feasible or near-optimal solution. Therefore, many researchers have proposed new solving methods called metaheuristic algorithms. These algorithms are used to solve complex optimization problems within reasonable computational time and cost.

Metaheuristics (M) are often algorithms using probability sampling. They attempt to find the global optimum (G) of a difficult optimization problem (with discontinuities "D," e.g.), without being trapped by local optimum (Figure 10.19).

Metaheuristics are formally defined as iterative stochastic algorithms, which attempt to converge to the best solution by sampling an objective function. They behave like search algorithms, attempting to learn the characteristics of the problem in order to find the best solution.

The basic concept of most metaheuristic algorithms is inspired by nature, animal behavior, or physical phenomena [84,85]. Metaheuristic algorithms are classified into three main classes: evolutionary algorithm, Swarm intelligence algorithm, and physics algorithm. The classification of the metaheuristic algorithms is illustrated in Figure 10.20. Evolutionary methods mimic the evolutionary process in nature to perform optimization. The second class of metaheuristic algorithms are Swarm-based methods, which mimic the behavior of animals in a group. Whereas physics-based methods perform optimization using the rules of physics in the universe.

10.3.3.1 The metaheuristics algorithms for DC–DC converters

Metaheuristic algorithms have been widely used in the field of renewable energy to find and track the maximum power point of the PV or wind system due to the

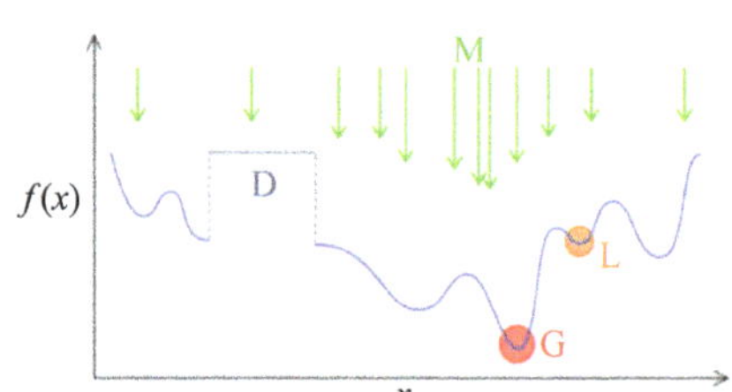

Figure 10.19 Local optimum and global optimum in the search space of $f(x)$

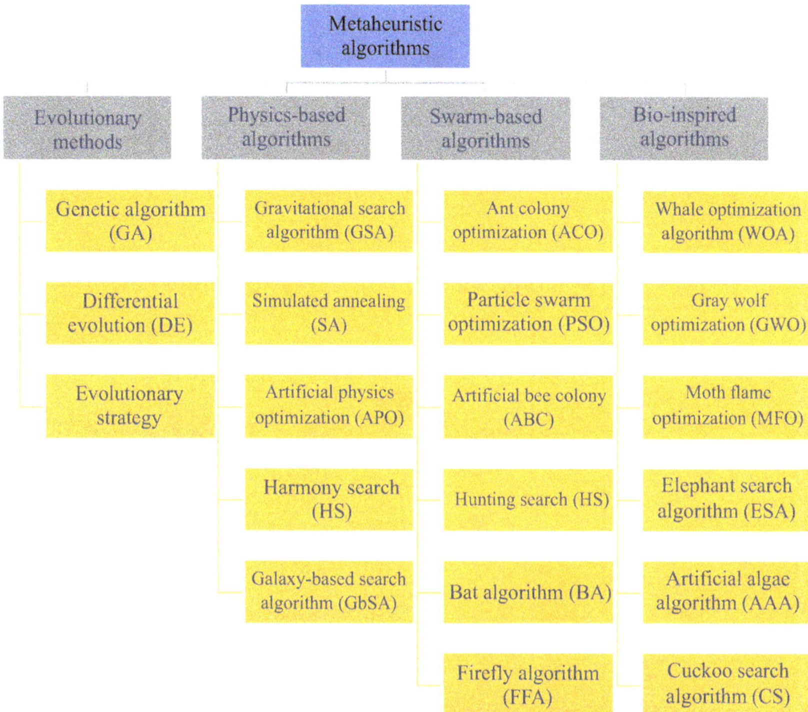

Figure 10.20 Classification of the metaheuristic algorithm

accuracy of these algorithms and their ability to adapt to external disturbances. For this, Table 10.10 summarizes the performances obtained for the different MPPT metaheuristic algorithms applied to the DC–DC converters.

10.3.3.2 The metaheuristics algorithms for DC–AC inverters

Concerning DC–AC inverters, metaheuristic algorithms have been used as a tool for the automatic and optimal adjustment of PI regulator coefficients. In [99], four algorithms i.e. imperialist competitive algorithm (ICA), genetic algorithm (GA), particle swarm optimization (PSO), and selfish herd optimization were used to optimally tune the coefficients of the PID regulator in order to stabilize the frequency. In [100], Whales optimization algorithm was used for the automatic adjustment of the coefficients of the PI regulator in order to control the currents injected into the grid. In [101], the authors presented a comparative study of four algorithms used for the optimal adjustment of the coefficients of the PI regulator in order to improve the quality of the currents injected into the grid.

Table 10.11 shows the performances obtained by the different metaheuristic's algorithms applied to the inverters operating in grid connected or in standalone mode. The table also shows the objectives of the command as well as the THD of the current and the voltage obtained.

Table 10.10 Performances obtained by MPPT metaheuristic algorithms

Metaheuristic algorithms	Converter type	Application	Modulation	Switching frequency	Sim/ exp	Tracking accuracy	Tracking speed	Reference
Modified GA	Buck	PV	PWM	NS	Exp	97%	0.83 s	[86]
DynNP-DE	Boost	PV	PWM	20 kHz	Sim	95%	0.95s	[87]
SA	Boost	PV	NS	NS	Exp	92%	NT	[88]
WOA	Boost	PV	NS	NS	Sim	99.97%	6.6s	[89]
LPSO	Boost	PV	PWM	10 kHz	Exp	99.98	0.45s	[90]
Modified GWO	SEPIC	PV	PWM	40 kHz	Sim	98.52%	NT	[91]
ABC	Boost	PV	PWM	25 kHz	Exp	98.41%	0.16s	[92]
Modified FA	Boost	PV	PWM	25 kHz	Sim	99.98%	2.2s	[93]
CS	Boost	PV	PWM	NS	Sim	NS	0.3s	[94]
FSSO	Quasi-Z-source	PV	PWM	10 kHz	Exp	99.93%	0.75s	[95]
ACO	NS	PV	PWM	NS	Exp	NS	0.4s	[96]
ABC	Zeta	PV	PWM	40 MHz	Sim	99.97%	0.1s	[97]
Modified ABC	Buck–boost	PV	PWM	20 kHz	Sim	99.87%	0.4s	[98]

NT, not tested; NS, not specified.

Table 10.11 Performances obtained for different metaheuristic algorithms applied to DC–AC inverters

Application	Inverter topology	Algorithms	Modulation	Switching frequency	Sim/exp	THD of V_o	THD of I_o	Control objective	Response for sudden changes
Stand-alone	H-bridge (3-ph)	ICA [99]	SVPWM	NS	Exp	NS	NS	-Voltage regulation for different type of load -Frequency regulation	+
		PSO [99]	SVPWM	NS	Exp	NS	NS	-Voltage regulation for different type of load -Frequency regulation	++
		GA [99]	SVPWM	NS	Exp	NS	NS	-Voltage regulation for different type of load -Frequency regulation	+
Grid-connected	NPC(3-ph)	PSO [102]	SVPWM	5 kHz	Sim	NS	NS	-Dc voltage regulation -Current regulation -THD minimization	+
	H-bridge (3-ph)	ALO [101]	NS	NS	Sim	NS	1.5%	-Current regulation -Dc voltage regulation	NS
		MFO [101]	NS	NS	Sim	NS	1.5%	-Current regulation -DC voltage regulation	NS
		WOA [100]	SVPWM	10 kHz	Sim	NS	0.09%	-Current regulation	NS
		GWO [101]	NS	NS	Sim	NS	2.1%	-Current regulation -Dc voltage regulation	NS

Good, +; very good, ++; NT, not tested; NS, not specified.

10.4 Comparative performance analysis

The comparison of the algorithms takes into account different parameters, such as the complexity of the command, the robustness to disturbance, implementation, and parameter dependency. The performance of each control strategy is shown in Table 10.12.

The choice of the converter and the control strategy essentially depends on the desired performance and the application in which the converter will be integrated. Generally, there is no unified approach for the choice of the control strategy, for this, we propose through the flowchart presented in Figure 10.21, an approach based on three criteria:

Table 10.12 Comparison of different control methods

Control approaches		Complexity of the command	Implementation	Robustness to disturbance	Parameter dependency
Linear controller	PI	Low	A, D	Low	High
	PR	Low	A, D	Low	High
Nonlinear controller	SMC	Moderate	A, D	Good	Low
	MPC	Moderate	D	Medium	Medium
Intelligent controller	FLC	Moderate	D	Medium	Low
	NNC	Moderate	D	Good	Low
	GA	Difficult	D	Good	Low
	PSO	Difficult	D	Good	Low
	ACO	Difficult	D	Good	Low

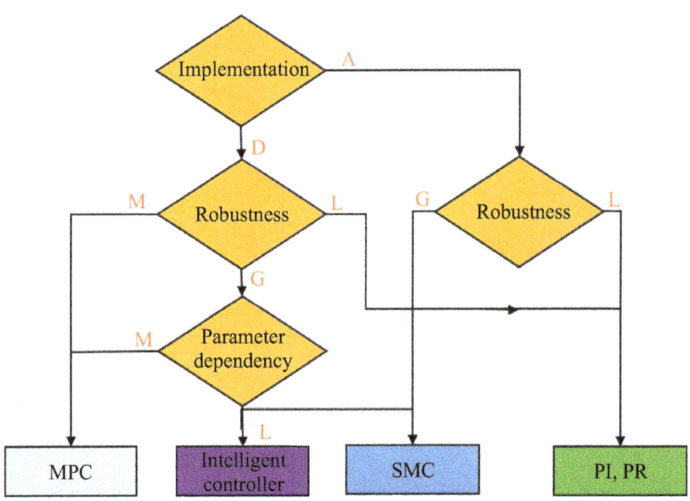

Figure 10.21 Flowchart of choice of a control algorithm

- Nature of the implementation circuit:
  - Analog (A)
  - or Digital (D)
- Desired robustness to disturbance:
  - High (H) for applications that require great robustness such as power converters ensuring the MPPT function where the input voltage is variable.
  - Medium (M) or low (L) for general applications where input voltage and load are less variable.
- Parameter dependency: The parameters of the controller depend on the mathematical model of the converter. Changing the internal parameters of the converter leads to performance degradation. So, for converters whose parameters are variable, it will be necessary to choose an algorithm whose dependence is low.

10.5 Conclusion

This last chapter of this book presented an overview of the control laws for power converters. Two types of power converters have been considered in this book: DC–DC converters with its different classes (isolated and non-isolated converters) and DC–DC inverters with its different classes (two-level and multi-level inverters). The variety of existing control methods has been classified into three classes: linear controllers, nonlinear controllers, and intelligent controllers. A comparison has been given in this chapter in order to select an appropriate control algorithm for a particular power converter. This choice essentially depends on the application in which the converter will be integrated. For example, when an inverter is operating in grid-tied mode, the conventional linear controller is sufficiently capable for following the reference. However, for standalone mode, the load is unknown, SM controller or intelligent controllers are more suitable to perform well under wide range disturbances and uncertainties.

References

[1] Utkin VI. Scope of the theory of sliding modes. In: Utkin VI, editor. *Sliding Modes in Control and Optimization*. Berlin, Heidelberg: Springer Berlin Heidelberg; 1992. p. 1–11. Available from: https://doi.org/10.1007/978-3-642-84379-2_1

[2] Tokat S, Fadali MS, and Eray O. A classification and overview of sliding mode controller sliding surface design methods. In: Yu X, Önder Efe M, editors. *Recent Advances in Sliding Modes: From Control to Intelligent Mechatronics*. Cham: Springer International Publishing; 2015. p. 417–39. Available from: https://doi.org/10.1007/978-3-319-18290-2_20

[3] Gorugantu R and Gandhi P. Control design based on novel parabolic sliding surfaces. In: *2012 12th International Workshop on Variable Structure Systems*. 2012. p. 410–5.

[4] Wu L, Liu J, Vazquez S, and Mazumder SK. Sliding mode control in power converters and drives: a review. *IEEE/CAA Journal of Automatica Sinica.* 2022;9(3):392–406.

[5] Komurcugil H and Biricik S. Time-varying and constant switching frequency-based sliding-mode control methods for transformerless DVR employing half-bridge VSI. *IEEE Transactions on Industrial Electronics.* 2017;64(4):2570–9.

[6] Bagheri F, Komurcugil H, and Kukrer O. Fixed switching frequency sliding-mode control methodology for single-phase LCL-filtered quasi-Z-source grid-tied inverters. In: *2018 IEEE 12th International Conference on Compatibility, Power Electronics and Power Engineering (CPE-POWERENG 2018)*. 2018. p. 1–6.

[7] Komurcugil H, Biricik S, Bayhan S, and Zhang Z. Sliding mode control: overview of its applications in power converters. *IEEE Industrial Electronics Magazine*. 2021;15(1):40–9.

[8] Tan SC, Lai YM, Cheung MKH, and Tse CK. On the practical design of a sliding mode voltage controlled buck converter. *IEEE Transactions on Power Electronics.* 2005;20(2):425–37.

[9] Tan SC, Lai YM, Tse CK, and Cheung MKH. A fixed-frequency pulsewidth modulation based quasi-sliding-mode controller for buck converters. *IEEE Transactions on Power Electronics.* 2005;20(6):1379–92.

[10] Komurcugil H. Adaptive terminal sliding-mode control strategy for DC–DC buck converters. *ISA Transactions.* 2012;51(6):673–81. Available from: https://www.sciencedirect.com/science/article/pii/S0019057812001164

[11] Leon-Masich A, Valderrama-Blavi H, Bosque-Moncusí JM, Maixé-Altés J, and Martínez-Salamero L. Sliding-mode-control-based boost converter for high-voltage–low-power applications. *IEEE Transactions on Industrial Electronics.* 2015;62(1):229–37.

[12] Tan SC, Lai YM, and Tse CK. A unified approach to the design of PWM-based sliding-mode voltage controllers for basic DC-DC converters in continuous conduction mode. *IEEE Transactions on Circuits and Systems I: Regular Papers.* 2006;53(8):1816–27.

[13] Yazici İ and Yaylaci EK. Fast and robust voltage control of DC–DC boost converter by using fast terminal sliding mode controller. *IET Power Electronics.* 2016;9(1):120–5. Available from: https://doi.org/10.1049/iet-pel.2015.0008

[14] Salimi M, Soltani J, Zakipour A, and Abjadi NR. Hyper-plane sliding mode control of the DC–DC buck/boost converter in continuous and discontinuous conduction modes of operation. *IET Power Electronics.* 2015;8(8):1473–82. Available from: https://doi.org/10.1049/iet-pel.2014.0578

[15] Komurcugil H, Biricik S, and Guler N. Indirect sliding mode control for DC–DC SEPIC converters. *IEEE Transactions on Industrial Informatics.* 2020;16(6):4099–108.

[16] Malesani L, Spiazzi RG, and Tenti P. Performance optimization of Cuk converters by sliding-mode control. *IEEE Transactions on Power Electronics.* 1995;10(3):302–9.

[17] Hasanah RN, Ardhenta L, Nurwati T, *et al.* Design of PI sliding mode control for Zeta DC-DC converter in PV system. *Bulletin of the Polish Academy of Sciences Technical Sciences.* 2022;70(3):140952.

[18] Salimi M, Soltani J, Zakipour A, and Hajbani V. Sliding mode control of the DC-DC flyback converter with zero steady-state error. In: *4th Annual International Power Electronics, Drive Systems and Technologies Conference*. 2013. p. 158–63.

[19] Baby A and Nithya M. Sliding mode controlled forward converter. In: *2015 IEEE 9th International Conference on Intelligent Systems and Control (ISCO)*. 2015. p. 1–4.

[20] Zhang H, Ma R, Han C, Xie R, Liang B, and Li Y. Advanced control design of interleaved boost converter for fuel cell applications. In: *IECON 2020 The 46th Annual Conference of the IEEE Industrial Electronics Society*. 2020. p. 5000–5.

[21] Kukrer O, Komurcugil H, and Doganalp A. A three-level hysteresis function approach to the sliding-mode control of single-phase UPS inverters. *IEEE Transactions on Industrial Electronics.* 2009;56(9):3477–86.

[22] Abrishamifar A, Ahmad A, and Mohamadian M. Fixed switching frequency sliding mode control for single-phase unipolar inverters. *IEEE Transactions on Power Electronics.* 2012;27(5):2507–14.

[23] Pichan M and Rastegar H. Sliding-mode control of four-leg inverter with fixed switching frequency for uninterruptible power supply applications. *IEEE Transactions on Industrial Electronics.* 2017;64(8):6805–14.

[24] Hao X, Yang X, Liu T, Huang L, and Chen W. A sliding-mode controller with multiresonant sliding surface for single-phase grid-connected VSI with an LCL filter. *IEEE Transactions on Power Electronics.* 2013;28 (5):2259–68.

[25] Guzman R, Vicuña LG de, Castilla M, Miret J, and Hoz J de la. Variable structure control for three-phase LCL-filtered inverters using a reduced converter model. *IEEE Transactions on Industrial Electronics.* 2018;65 (1):5–15.

[26] Sebaaly F, Vahedi H, Kanaan HY, Moubayed N, and Al-Haddad K. Design and implementation of space vector modulation-based sliding mode control for grid-connected 3L-NPC inverter. *IEEE Transactions on Industrial Electronics.* 2016;63(12):7854–63.

[27] Zakipour A, Shokri Kojori S, and Tavakoli Bina M. Closed-loop control of the grid-connected Z-source inverter using hyper-plane MIMO sliding mode. *IET Power Electronics.* 2017;10(15):2229–41. Available from: https://doi.org/10.1049/iet-pel.2017.0076

[28] Wang F, Mei X, Rodriguez J, and Kennel R. Model predictive control for electrical drive systems-an overview. *CES Transactions on Electrical Machines and Systems.* 2017;1(3):219–30.

[29] Vazquez S, Rodriguez J, Rivera M, Franquelo LG, and Norambuena M. Model predictive control for power converters and drives: advances and trends. *IEEE Transactions on Industrial Electronics.* 2017;64(2):935–47.

[30] Andrés-Martínez O, Flores-Tlacuahuac A, Ruiz-Martinez OF, and Mayo-Maldonado JC. Nonlinear model predictive stabilization of DC–DC boost converters with constant power loads. *IEEE Journal of Emerging and Selected Topics in Power Electronics.* 2021;9(1):822–30.

[31] Chen J, Chen Y, Tong L, and Peng L. Robust explicit model predictive control based on state feedback linearization for buck converter. In: *2020 IEEE 9th International Power Electronics and Motion Control Conference (IPEMC2020-ECCE Asia).* 2020. p. 3156–60.

[32] Jeong M, Fuchs S, and Biela J. When FPGAs meet regionless explicit MPC: an implementation of long-horizon linear MPC for power electronic systems. In: *IECON 2020 The 46th Annual Conference of the IEEE Industrial Electronics Society*. 2020. p. 3085–92.

[33] Wang B, Xian L, and Tseng KJ. Dynamic model predictive voltage control for single-input dual-output flyback converter. In: *2016 IEEE 11th Conference on Industrial Electronics and Applications (ICIEA).* 2016. p. 1212–7.

[34] Guler N, Biricik S, Bayhan S, and Komurcugil H. Model predictive control of DC–DC SEPIC converters with autotuning weighting factor. *IEEE Transactions on Industrial Electronics.* 2021;68(10):9433–43.

[35] Cunha RBA, Inomoto RS, Altuna JAT, Costa FF, di Santo SG, and Sguarezi Filho AJ. Constant switching frequency finite control set model predictive control applied to the boost converter of a photovoltaic system. *Solar Energy.* 2019;189:57–66. Available from: https://www.sciencedirect.com/science/article/pii/S0038092X19306838

[36] Albira ME and Zohdy MA. Adaptive model predictive control for DC-DC power converters with parameters' uncertainties. *IEEE Access.* 2021;9:135121–31.

[37] Neely J, DeCarlo R, and Pekarek S. Real-time model predictive control of the Ćuk converter. In: *2010 IEEE 12th Workshop on Control and Modeling for Power Electronics (COMPEL).* 2010. p. 1–8.

[38] Shen Y, Xie L, and Li X. Explicit hybrid model predictive control of the forward DC-DC converter. In: *2013 25th Chinese Control and Decision Conference (CCDC).* 2013. p. 638–42.

[39] Tarisciotti L, Chen L, Shao S, Dragičević T, Wheeler P, and Zanchetta P. Finite control set model predictive control for dual active bridge converter. *IEEE Transactions on Industry Applications.* 2022;58(2):2155–65.

[40] Talbi B, Krim F, Laib A, and Sahli A. Model predictive voltage control of a single-phase inverter with output LC filter for stand-alone renewable energy systems. *Electrical Engineering.* 2020;102(3):1073–82. Available from: https://doi.org/10.1007/s00202-020-00936-5

[41] Lezana P, Aguilera R, and Quevedo DE. Model predictive control of an asymmetric flying capacitor converter. *IEEE Transactions on Industrial Electronics.* 2009;56(6):1839–46.

[42] Chen H, Wang D, Tang S, Yin X, Wang J, and Shen ZJ. Continuous control set model predictive control for three-level flying capacitor boost converter with constant switching frequency. *IEEE Journal of Emerging and Selected Topics in Power Electronics.* 2021;9(5):5996–6007.

[43] Danayiyen Y, Lee K, Choi M, and Lee Y il. Model predictive control of uninterruptible power supply with robust disturbance observer. *Energies (Basel).* 2019;12(15). Available from: https://www.mdpi.com/1996-1073/12/15/2871

[44] Cortes P, Ortiz G, Yuz JI, Rodriguez J, Vazquez S, and Franquelo LG. Model predictive control of an inverter with output LC filter for UPS applications. *IEEE Transactions on Industrial Electronics.* 2009;56(6):1875–83.

[45] Vazquez S, Aguilera RP, Acuna P, *et al.* Model predictive control for single-phase NPC converters based on optimal switching sequences. *IEEE Transactions on Industrial Electronics.* 2016;63(12):7533–41.

[46] Guzman R, Vicuña LG de, Camacho A, Miret J, and Rey JM. Receding-horizon model-predictive control for a three-phase VSI with an LCL filter. *IEEE Transactions on Industrial Electronics.* 2019;66(9):6671–80.

[47] Kang L, Cheng J, Hu B, Luo X, and Zhang J. A simplified optimal-switching-sequence MPC with finite-control-set moving horizon optimization for grid-connected inverter. *Electronics (Basel).* 2019;8(4). Available from: https://www.mdpi.com/2079-9292/8/4/457

[48] Taheri A and Zhalebaghi MH. A new model predictive control algorithm by reducing the computing time of cost function minimization for NPC inverter in three-phase power grids. *ISA Transactions.* 2017;71:391–402. Available from: https://www.sciencedirect.com/science/article/pii/S0019057817305128

[49] Han J, Liu L, Yao G, and Tang T. Finite-control-set model predictive control for asymmetrical cascaded H-bridge multilevel grid-connected inverter with flying capacitor. *IEEJ Transactions on Electrical and Electronic Engineering.* 2021;16(10):1328–35. Available from: https://onlinelibrary.wiley.com/doi/abs/10.1002/tee.23440

[50] Ramírez RO, Baier CR, Espinoza J, and Villarroel F. Finite control set MPC with fixed switching frequency applied to a grid connected single-phase cascade H-bridge inverter. *Energies (Basel).* 2020;13(20):5475. Available from: https://www.mdpi.com/1996-1073/13/20/5475

[51] Ying H. Basic fuzzy mathematics for fuzzy control and modeling. In: *Fuzzy Control and Modeling: Analytical Foundations and Applications*. 2000. p. 1–14.

[52] Zadeh LA, Klir GJ, and Yuan B. *Fuzzy Sets, Fuzzy Logic, and Fuzzy Systems*. World Scientific; 1996. Available from: https://www.worldscientific.com/doi/abs/10.1142/2895

[53] Nayak JR, Shaw B, and Sahu BK. Application of adaptive-SOS (ASOS) algorithm based interval type-2 fuzzy-PID controller with derivative filter for automatic generation control of an interconnected power system. *Engineering Science and Technology, an International Journal.* 2018;21(3):465–85. Available from: https://www.sciencedirect.com/science/article/pii/S2215098617315811

[54] Gupta T, Boudreaux RR, Nelms RM, and Hung JY. Implementation of a fuzzy controller for DC-DC converters using an inexpensive 8-b microcontroller. *IEEE Transactions on Industrial Electronics.* 1997;44(5):661–9.
[55] Mattavelli P, Rossetto L, Spiazzi G, and Tenti P. General-purpose fuzzy controller for DC-DC converters. *IEEE Transactions on Power Electronics.* 1997;12(1):79–86.
[56] Atacak I and Bay OF. A type-2 fuzzy logic controller design for buck and boost DC–DC converters. *Journal of Intelligent Manufacturing.* 2012;23(4):1023–34. Available from: https://doi.org/10.1007/s10845-010-0388-1
[57] Elkhateb A, Rahim NA, Selvaraj J, and Uddin MN. Fuzzy-logic-controller-based SEPIC converter for maximum power point tracking. *IEEE Transactions on Industry Applications.* 2014;50(4):2349–58.
[58] Radjai T, Gaubert JP, Rahmani L, and Mekhilef S. Experimental verification of P&O MPPT algorithm with direct control based on Fuzzy logic control using CUK converter. *International Transactions on Electrical Energy Systems*. 2015;25(12):3492–508. Available from: https://onlinelibrary.wiley.com/doi/abs/10.1002/etep.2047
[59] Shahid MA, Abbas G, Hussain MR, *et al.* Artificial intelligence-based controller for DC-DC flyback converter. *Applied Sciences.* 2019;9(23):5108–122. Available from: https://www.mdpi.com/2076-3417/9/23/5108
[60] Buccella C, Cecati C, Latafat H, and Razi K. Comparative transient response analysis of LLC resonant converter controlled by adaptive PID and fuzzy logic controllers. In: *IECON2012 – 38th Annual Conference on IEEE Industrial Electronics Society*. 2012. p. 4729–34.
[61] Tiwary N, Naik VN, Panda AK, Narendra A, and Lenka RK. Fuzzy logic based direct power control of dual active bridge converter. In: *2021 1st International Conference on Power Electronics and Energy (ICPEE)*. 2021. p. 1–5.
[62] Zheng X, Zaman H, Wu X, Ali H, and Khan S. Direct fuzzy logic controller for voltage control of standalone three phase inverter. In: *2017 International Electrical Engineering Congress (iEECON)*. 2017. p. 1–4.
[63] Ismailou AA, Wang H, Lu T, Wang C, and Shi F. MPPT for single phase five level Quasi-z-source photovoltaic inverter with fuzzy controller. *IOP Conference Series: Earth and Environmental Science.* 2020;453(1):12043. Available from: https://doi.org/10.1088/1755-1315/453/1/012043
[64] Cecati C, Ciancetta F, and Siano P. A multilevel inverter for photovoltaic systems with fuzzy logic control. *IEEE Transactions on Industrial Electronics.* 2010;57(12):4115–25.
[65] Altin N and Ozdemir S. Three-phase three-level grid interactive inverter with fuzzy logic based maximum power point tracking controller. *Energy Conversion and Management.* 2013;69:17–26. Available from: https://www.sciencedirect.com/science/article/pii/S0196890413000320
[66] Hannan MA, Ghani ZA, Mohamed A, and Uddin MN. Real-time testing of a fuzzy logic controller based grid-connected photovoltaic inverter system. In: *2014 IEEE Industry Application Society Annual Meeting*. 2014. p. 1–8.

[67] Bose BK. Neural network applications in power electronics and motor drives—an introduction and perspective. *IEEE Transactions on Industrial Electronics*. 2007;54(1):14–33.

[68] Kumar P, Lai SH, Wong JK, *et al.* Review of nitrogen compounds prediction in water bodies using artificial neural networks and other models. *Sustainability*. 2020;12(11):1–26. Available from: https://www.mdpi.com/2071-1050/12/11/4359

[69] Dong W, Li S, Fu X, Li Z, Fairbank M, and Gao Y. Control of a buck DC/DC converter using approximate dynamic programming and artificial neural networks. *IEEE Transactions on Circuits and Systems I: Regular Papers*. 2021;68(4):1760–8.

[70] Khan HS, Mohamed IS, Kauhaniemi K, and Liu L. Artificial neural network-based voltage control of DC/DC converter for DC microgrid applications. In: *2021 6th IEEE Workshop on the Electronic Grid (eGRID)*. 2021. p. 1–6.

[71] Farh HMH, Eltamaly AM, Ibrahim AB, Othman MF, and Al-Saud MS. Dynamic global power extraction from partially shaded photovoltaic using deep recurrent neural network and improved PSO techniques. *International Transactions on Electrical Energy Systems*. 2019;29(9):e12061. Available from: https://onlinelibrary.wiley.com/doi/abs/10.1002/2050-7038.12061

[72] Sundaramoorthy S, Umamaheswari MG, Marimuthu G, and Lekshmisree B. Hopfield neural network-based average current mode control of synchronous SEPIC converter. *IETE Journal of Research*. 2021;0(0):1–19. Available from: https://doi.org/10.1080/03772063.2021.1926344

[73] Shahid MA, Abbas G, Hussain MR, *et al.* Artificial intelligence-based controller for DC-DC flyback converter. *Applied Sciences*. 2019;9(23):5108. Available from: https://www.mdpi.com/2076-3417/9/23/5108

[74] Utomo WM, Bakar A, Ahmad M, Taufik T, and Heriansyah R. Online learning neural network control of buck-boost converter. In: *2011 Eighth International Conference on Information Technology: New Generations*. 2011. p. 485–9.

[75] Mahdavi J, Nasiri MR, Agah A, and Emadi A. Application of neural networks and state-space averaging to DC/DC PWM converters in sliding-mode operation. *IEEE/ASME Transactions on Mechatronics*. 2005;10(1):60–7.

[76] Utomo WM, Yi SS, Buswig YMY, Haron ZA, Bakar AA, and Ahmad MZ. Voltage tracking of a DC-DC flyback converter using neural network control. *International Journal of Power Electronics and Drive System (IJPEDS)*. 2012;2(1):35–42.

[77] Tang Y, Hu W, Xiao J, *et al.* RL-ANN-based minimum-current-stress scheme for the dual-active-bridge converter with triple-phase-shift control. *IEEE Journal of Emerging and Selected Topics in Power Electronics*. 2022;10(1):673–89.

[78] Mohamed IS, Rovetta S, Do TD, Dragicević T, and Diab AAZ. A neural-network-based model predictive control of three-phase inverter with an output LC filter. *IEEE Access*. 2019;7:124737–49.

[79] Novak M and Blaabjerg F. Supervised imitation learning of FS-MPC algorithm for multilevel converters. In: *2021 23rd European Conference on Power Electronics and Applications (EPE'21 ECCE Europe)*. 2021. p. P.1–P.10.

[80] Jamma M, Bennassar A, Barara M, and Akherraz M. Advanced direct power control for grid-connected distribution generation system based on fuzzy logic and artificial neural networks techniques. *International Journal of Power Electronics and Drive Systems (IJPEDS)*. 2017;8(3):979.

[81] Bakeer A, Mohamed IS, Malidarreh PB, Hattabi I, and Liu L. An artificial neural network-based model predictive control for three-phase flying capacitor multilevel inverter. *IEEE Access*. 2022;10:70305–16.

[82] Babaie M, Sharifzadeh M, and Al-Haddad K. Three-phase grid-connected NPC inverter based on a robust artificial neural network controller. In: *2020 IEEE Power & Energy Society General Meeting (PESGM)*. 2020. p. 1–5.

[83] Fu X and Li S. A novel neural network vector control for single-phase grid-connected converters with L, LC and LCL filters. *Energies (Basel)*. 2016;9(5):328. Available from: https://www.mdpi.com/1996-1073/9/5/328

[84] Minai AF and Malik H. Metaheuristics paradigms for renewable energy systems: advances in optimization algorithms. Metaheuristic and evolutionary computation: algorithms and applications. In: *Studies in Computational Intelligence*, vol. 916. Singapore: Springer; 2021. https://doi.org/10.1007/978-981-15-7571-6_2

[85] Malik H, Iqbal A, Joshi P, Agrawal S, and Bakhsh FI. Metaheuristic and evolutionary computation: algorithms and applications. In: *Studies in Computational Intelligence*, vol. 916. Singapore: Springer; 2021. https://doi.org/10.1007/978-981-15-7571-6

[86] Daraban S, Petreus D, and Morel C. A novel MPPT (maximum power point tracking) algorithm based on a modified genetic algorithm specialized on tracking the global maximum power point in photovoltaic systems affected by partial shading. *Energy*. 2014;74:374–88. Available from: https://www.sciencedirect.com/science/article/pii/S0360544214008184

[87] Tajuddin MFN, Ayob SM, and Salam Z. Global maximum power point tracking of PV system using dynamic population size differential evolution (DynNP-DE) algorithm. In: *2014 IEEE Conference on Energy Conversion (CENCON)*. 2014. p. 254–9.

[88] Lyden S and MdE H. A simulated annealing global maximum power point tracking approach for PV modules under partial shading conditions. *IEEE Transactions on Power Electronics*. 2016;31(6):4171–81.

[89] Santhan Kumar CH and Srinivasa Rao R. A novel global MPP tracking of photovoltaic system based on whale optimization algorithm. *International Journal of Renewable Energy Development*. 2016;5(3):225–32.

[90] Prasanth Ram J and Rajasekar N. A new robust, mutated and fast tracking LPSO method for solar PV maximum power point tracking under partial shaded conditions. *Applied Energy*. 2017;201:45–59. Available from: https://www.sciencedirect.com/science/article/pii/S0306261917306049

[91] Hasan FR, Prasetyono E, and Sunarno E. A modified maximum power point tracking algorithm using grey wolf optimization for constant power generation of photovoltaic system. In: *2021 International Conference on Artificial Intelligence and Mechatronics Systems (AIMS)*. 2021. p. 1–6.

[92] González-Castaño C, Restrepo C, Kouro S, and Rodriguez J. MPPT algorithm based on artificial bee colony for PV system. *IEEE Access*. 2021;9:43121–33.

[93] Farzaneh J, Keypour R, and Khanesar MA. A new maximum power point tracking based on modified firefly algorithm for PV system under partial shading conditions. *Technology and Economics of Smart Grids and Sustainable Energy*. 2018;3(1):9. Available from: https://doi.org/10.1007/s40866-018-0048-7

[94] Raj A and Gupta M. Numerical simulation and comparative assessment of improved cuckoo search and PSO based MPPT system for solar photovoltaic system under partial shading condition. *Turkish Journal of Computer and Mathematics Education*. 2021;12:3842.

[95] Singh N, Gupta KK, Jain SK, Dewangan NK, and Bhatnagar P. A flying squirrel search optimization for MPPT under partial shaded photovoltaic system. *IEEE Journal of Emerging and Selected Topics in Power Electronics*. 2021;9(4):4963–78.

[96] Jiang LL, Maskell DL, and Patra JC. A novel ant colony optimization-based maximum power point tracking for photovoltaic systems under partially shaded conditions. *Energy and Buildings*. 2013;58:227–36. Available from: https://www.sciencedirect.com/science/article/pii/S0378778812006366

[97] Fanani MR, Sudiharto I, and Ferdiansyah I. Implementation of maximum power point tracking on PV system using artificial bee colony algorithm. In: *2020 3rd International Seminar on Research of Information Technology and Intelligent Systems (ISRITI)*. 2020. p. 117–22.

[98] Li N, Mingxuan M, Yihao W, Lichuang C, Lin Z, and Qianjin Z. Maximum power point tracking control based on modified ABC algorithm for shaded PV system. In: *2019 AEIT International Conference of Electrical and Electronic Technologies for Automotive (AEIT AUTOMOTIVE)*. 2019. p. 1–5.

[99] Rajamand S. Effective control of voltage and frequency in microgrid using adjustment of PID coefficients by metaheuristic algorithms. *IETE Journal of Research*. 2021;0(0):1–14. Available from: https://doi.org/10.1080/03772063.2020.1769509

[100] Qazi SH, bin Mustafa MW, Soomro S, and Larik RM. An optimal current controller for photovoltaic system based three phase grid using whales optimization algorithm. In: *2017 IEEE Conference on Energy Conversion (CENCON)*. 2017. p. 15–20.

[101] Aouchiche N. Meta-heuristic optimization algorithms based direct current and DC link voltage controllers for three-phase grid connected photovoltaic inverter. *Solar Energy*. 2020;207:683–92. Available from: https://www.sciencedirect.com/science/article/pii/S0038092X20307003

[102] Thameur A, Noureddine B, Abdelhalim B, *et al.* Particle swarm optimization of PI controllers in grid-connected PV conversion cascade based three levels NPC inverter. In: *2020 IEEE International Conference on Environment and Electrical Engineering and 2020 IEEE Industrial and Commercial Power Systems Europe (EEEIC/I&CPS Europe)*. 2020. p. 1–5.

Index

www.ingramcontent.com/pod-product-compliance
Lightning Source LLC
Chambersburg PA
CBHW050514190326
41458CB00005B/1536
9781839537400